Inno'va-tion

Inno'va-tion

Essays by Leading Canadian Researchers

Edited by James Downey and Lois Claxton

KEY PORTER BOOKS

National Library of Canada Cataloguing in Publication Data

Innovation: essays by leading Canadian researchers / editors,
James Downey and Lois Claxton.

Includes index.
ISBN 1-55263-500-7

1. Research—Canada. 2. Technological innovations—Canada.
I. Downey, James, 1939- . II. Claxton, Lois.

Q180.C3I55 2002 507'.2071 C2002-903848-0

The publisher gratefully acknowledges the support of the Canada Council for the Arts and the Ontario Arts Council for its publishing program.

We acknowledge the financial support of the Government of Canada through the Book Publishing Industry Development Program (BPIDP) for our publishing activities.

Key Porter Books Limited
70 The Esplanade
Toronto, Ontario
Canada M5E 1R2

www.keyporter.com

Cover design: Peter Maher
Electronic formatting: Heidy Lawrance Associates

Printed and bound in Canada

02 03 04 05 06 07 6 5 4 3 2 1

Table of Contents

Acknowledgements

Inno'va-tion was David Strangway's idea. As president of the Canada Foundation for Innovation and former president of two of Canada's finest research universities, he saw the need for a book that would give the informed but non-specialist reader a sense of the range, depth, and excitement of the scholarly inquiry taking place in our nation's laboratories and libraries. When asked if I would commission and edit a collection of essays by leading researchers, I accepted with enthusiasm.

Before inviting contributions, I enlisted the help of several people very familiar with the Canadian research scene to identify a representative sample of the best work being done. Thomas Brzustowski, President of the Natural Sciences and Engineering Research Council, Arthur Carty, President of the National Research Council, and Marc Renaud, President of the Social Sciences and Humanities Research Council, were especially helpful. Almost without exception, those invited to contribute essays agreed to do so. Of course it has not been possible to feature all, or even most, of the exciting fields of inquiry being explored, or more than a few of the many excellent researchers at work in Canada today. My sense is that it would be possible to publish several more collections of this kind without difficulty or duplication.

I should like to thank the authors of the essays. Getting to know something of them and their work has been personally rewarding. Linda Kenyon and Patricia Bow both lent helpful editorial assistance at an early stage, and at the end Patrick Mikhail brought his impressive professional skills to bear on editorial refinements of the text. Suzanne Corbeil of the Canada Foundation for Innovation was wonderfully supportive throughout. Most of all, I wish to thank my co-editor, Lois Claxton, for her superb management of the project.

James Downey

Foreword

Since it was established in 1997, the Canada Foundation for Innovation has been firmly focused on two key elements of research in Canada—infrastructure and partnerships. By strengthening the ability of our nation's researchers to carry out world-class research, by inspiring collaboration, and by promoting and building valuable partnerships with research institutions all across our country, the CFI is helping to ensure Canada's continued prosperity in the global economy.

Among the many valuable partners who contribute to the success of the CFI's programs and goals are Canada's research-funding organizations. Academic research in Canada has been sustained, enriched, and made vital to our nation's needs by the leadership of these organizations and the outstanding individuals who lead them: Alan Bernstein, President of the Canadian Institutes of Health Research; Thomas Brzustowski, President of the Natural Sciences and Engineering Research Council; Marc Renaud, President of the Social Sciences and Humanities Research Council; René Durocher, Executive Director of the Canada Research Chairs; and Martin Godbout, President and CEO of Genome Canada. We thank them—and the organizations they represent—for their commitment to research and innovation in Canada.

In recognizing those who create and advance innovation in Canada, we must reserve our greatest appreciation and admiration for our nation's researchers. They are at the core of our most valuable and essential partnerships—without them, there would be no innovation to speak of. Their excellence, imagination, and creativity are the truly impressive part of the innovation equation. Together, they are strengthening Canada's scientific leadership and demonstrating to the world's research community our commitment to building and innovating for the future.

The Canada Foundation for Innovation is pleased to present this book and the essays of the twenty-five talented Canadian researchers who have so generously and eloquently shared their ideas and experiences. We dedicate *Inno'va-tion* to them and the entire community of researchers all across Canada. They are prime examples of the people and projects that will have a profound impact on the overall quality of life of countless Canadians—for many years to come.

David Strangway
President and CEO
Canada Foundation for Innovation

Introduction

Since 1970, Canada has won only three Nobel Prizes in science. At least eight Canadians have won Nobel Prizes elsewhere during the same period, and several other winners spent part of their careers in Canada. There can be little doubt that, at the highest reaches of academic performance, Canada has suffered a serious brain drain. As a senior academic administrator for much of the past quarter century, I have seen close-up the gap between reach and grasp in Canada's research effort, a gap that has cost us dearly in national pride and perhaps in more practical ways as well.

It has been heartening, therefore, to witness the recent re-investment in research being made by our national government through the creation of the Canada Foundation for Innovation, to strengthen research infrastructure in universities and hospitals; the Canada Research Chairs program, to retain and attract outstanding researchers; and Genome Canada, to advance Canada's standing in the crucial area of genomics. In addition, there has been a restoration of budgets at our federal granting councils. The provinces as well, each in its own way, have sought to provide matching support, and Canadian industry and foundations have also caught the spirit. Altogether, billions of new dollars are being invested in Canadian research.

To make the desired difference, however, such a welcome commitment must be sustained. It must also be matched by a commitment to our universities to ensure that they can provide education comparable in quality to that available in other advanced countries. In the end, the future of Canada as a research nation will depend not only on what happens in our laboratories, but equally on what happens in our undergraduate classrooms, where intellectual curiosity is aroused and talent fostered.

Inno'va-tion gives unmistakable proof that a renewal of Canadian research is underway, one that is refurbishing and expanding the physical capabilities of Canada's research institutions while helping them to attract and retain a larger share of the brilliant and adventurous minds we have so readily exported in the past. This renewal is making Canada, after too long a period of under-investment, a more progressive and exciting place to pursue research.

Some conclusions are inescapable from these essays. First, traditional categorization (science, engineering, social science, medical science, humanities) doesn't work well for current research. Nearly all of the essays in this book cross disciplinary boundaries in one way or another. Second, information systems technology is increasingly a boundary-crossing feature of almost all research, from e-health to e-business to e-governance to e-learning. Inevitably, therefore, contemporary research is more than ever a collaborative enterprise. Little of value is accomplished in any field today that doesn't involve the convergence of expertise from a number of cognate disciplines. In this respect, Canadian research has been helped tremendously in recent years by the existence of the Canadian Institute for Advanced Research and the Centres of Excellence programs.

"Convergence" is the popular term for the blending of computing, telecommunications, and broadcast technologies. But it could equally apply to other knowledge mergings that are taking place. It applies to the work that Linda Hutcheon and colleagues are doing in the area of literary history by marshalling the expertise of geographers, sociologists, anthropologists, economists, linguists, and musicologists, as well as film and art historians, and, of course, literary critics. Convergence would also seem to fit the field of physics as described by Catherine Kallin: "In physics, as in many other occupations, effective collaboration is a proven way of getting results. Although there may still be room for a solitary Einstein to discover the laws underpinning the universe purely through the power of thought, more and more of what we know is being discovered by people working together. By combining their different talents and expertise, and stimulating each other to do the best they can." In the scientific research of the twenty-first century, the "Renaissance man" of past centuries is being replaced by the "Renaissance team" of highly talented men and women working together.

Similarly, in his essay, John Robinson not only describes a multidisciplinary approach to understanding the central issues of sustainability, but describes an even larger convergence: "...it is becoming clear that the border between the natural and the human is not a stable boundary. Rather, it is a shifting borderland, in which technological change, together with changes in our profoundest understandings of the social construction of culture and nature, are altering the ground on which we stand, both literally and metaphorically."

Nowhere is convergence more dramatic than in the area of genomic and genetic research, where scientific, medical, and moral choices have converged so profoundly that we have scarcely yet defined the issues, let alone resolved them, as several contributors to this collection describe.

While researchers may use highly sophisticated machines, they themselves are subject to the same vagaries of chance experience as the rest of us. Their discoveries are often a unique and unpredictable mix of curiosity, circumstance, skill, and personal reflection. In these essays, we are sometimes given a glimpse of how the professional and the personal connect. Sylvain Houle makes a poignant connection between the professional and the personal when he relates how he came to understand, through his research into the functioning of the human brain, that the seemingly bizarre behaviour of a university classmate at exam time many years ago was really the onset of schizophrenia.

Similarly, in describing the Canadian Families Project, Eric Sager reveals, in a very personal way, how present concerns about the family can influence what historians choose to study: "I found myself contributing to certain trends when my marriage ended and I joined the ranks of single parents. I began to read the literature on children of divorced parents, this time for more than academic interest. The personal, the political, and the intellectual had converged." What this interest led to in terms of analysis of census data is but one of the many counter-intuitive outcomes of the research described in this volume.

For all its justly celebrated "method," science is frequently influenced by the adventitious and the serendipitous. There is plenty of evidence here of carefully laid plans in one direction leading to a surprise discovery in another. Had Marc-André Sirard and his colleagues not needed to use heparin to prevent coagulation of blood in the cows from which they were trying to retrieve ova, they would not have discovered that this drug was useful as an agent to make sperm more fertile in vitro. It is often, as Martha Salcudean reminds us, the combination of scientific knowledge and human ingenuity that has led to most of the marvellous inventions that have blessed—and occasionally cursed—our lives.

A final recurring theme is gratitude. Again and again, authors credit mentors, colleagues, and students for contributing to the success they have achieved. In doing so, they remind us once more that research, today more than in the past, and in almost every field, is a team sport. Gratitude is also expressed repeatedly to the federal granting councils whose stewardship and leadership of Canadian research have been, and continue to be, simply outstanding.

"Why should we do all this?" Simon Lilly asks of astrophysicists' quest to voyage to the farthest reaches of the universe by means of very expensive telescopes. His answer goes to the very soul of academic research:

> Our goal is simple. We seek to understand humanity's place in the Universe, to understand our "origins." We seek to understand how we, and all that we see around us, came to be. Astronomers at the turn of the millennium have adopted the goal of gaining a causal understanding of the sequence of events that started with the Big Bang and ended with the

emergence of DNA. This is an ambitious goal, possibly a presumptuous one, but one with a certain nobility and grandeur. It is appropriate that Canadians are playing their part in what has become a truly global endeavour.

This may be star-gazing in the metaphorical as well as the literal sense, but it is important to remember that today's curiosity-driven research is tomorrow's technology, and we can never be sure where the next revolution in knowledge will come from. The digital world that now envelops us and occupies so much of our lives was, not so long ago, a series of seemingly harebrained ideas and experiments in the minds and laboratories of physicists, mathematicians, and engineers.

What this collection reveals, above all, is that Canada's research community is ready to do its part. It is brimming with intelligence, energy, and commitment, in touch with the vital issues of our time, and probing the deeper meanings of "innovation." Michael Wolfson uses "proprioception" (a term that refers to the set of neurons in the brain that tells us our body's position at any given time) in a metaphorical way to describe a country's statistical system. As he writes, "It continually provides us vital information on where we are, and where we are heading." The same metaphor might be applied to a nation's research capability: It gives us a sense of where we are and provides us with the clues to interpret choices about where we are heading.

Not every discovery, however, will be comforting. Some, like David Schindler's analysis of what we are doing to our immense freshwater resources, are alarming and disturbing. But this too, and perhaps especially, we need to know. Only sound research can reveal what is wrong with the way we are treating nature, and only the application of wisdom to that knowledge can show us how to mobilize the resources—material, intellectual, and moral—that are needed to correct it.

In the end, it is not the knowledge discoverers who will choose the future. Their job, as Fred Longstaffe writes, "is to continue to strive, to ask the right questions, to make the best experiments, and to follow them wherever they lead us." And as this collection shows, our knowledge discoverers are doing their job well. But it is we ourselves, through the social and political institutions we have created and through our individual choices as citizens and consumers, who must choose among the array of alternative futures that confront us. Learning more about what futures are possible is the beginning of the wisdom we will need in order to choose well. *Inno'va-tion* is an important contribution to the understanding of what our choices are.

James Downey
Waterloo, Ontario
September 2002

In September 1999, *Dr. Douglas Boyd* performed the world's first totally closed-chest, computer-enhanced, robot-assisted, beating-heart coronary bypass operation. Today, he continues his innovative work in London at the University of Western Ontario's National Centre of Advanced Surgery and Robotics—where he is both the centre's director and an associate professor of surgery.

Dr. Boyd's career in medicine began at the University of Ottawa, where he first did his undergraduate medical training. He followed this with general surgery and cardiothoracic surgical training at the University of Ottawa Heart Institute, where he did a fellowship in cardiac transplantation and mechanical-assist devices. Later, at the University of Toronto, he pursued a master's degree in medical education before heading to Washington University in St. Louis, Missouri, where he obtained additional experience in thoracic transplantation. He also trained for minimally invasive surgery at the Advanced Laparoscopy Training Center in Atlanta, Georgia, and the Minimally Invasive Surgical Training Institute in Baltimore, Maryland.

Dr. Boyd has received numerous surgical and research awards, including "Best Research Paper" at a meeting of the International Society of Minimally Invasive Surgery in Paris, France, in the fall of 1998.

Up Everest with a Scalpel

W. Douglas Boyd

Adrenaline was flowing freely on September 24, 1999—the day we climbed our Everest. We were in uncharted territory, all our senses heightened. Even though we were as prepared as we could possibly be, there was still some trepidation about the job ahead of us.

We had done endless hours of work in the laboratory, and operated on hundreds of other patients. We had become intimately familiar with our robotic assistant and its space-born capabilities. Nothing, however, could fully prepare us for the experience that was soon to come. We would operate on a beating heart from eight feet away—and we would do it with a robot holding the scalpel.

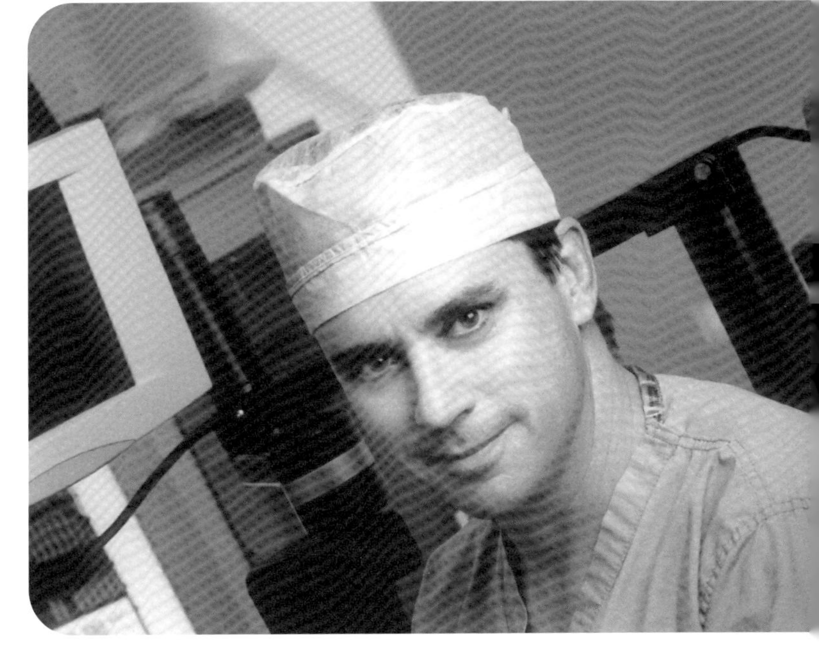

Six hours later, after performing the world's first robotic-assisted closed-chest bypass surgery on a beating heart, I felt both drained and elated. My colleagues and I at the London Health Sciences Centre in London, Ontario, had been working toward this moment for three years. We had set out to revolutionize the way heart surgery is performed, and we had done it.

The patient, a sixty-year-old dairy farmer, went home four days later with only a few small incisions to show for his part in the making of medical history. Three weeks after the operation he was back at work. That success, not our work in the operating room, was the real cause for celebration.

Anyone who has witnessed the effects first-hand on a loved one knows that conventional cardiac surgery delivers a ferocious blow to the body. You start by making a foot-long incision in the chest, splitting the breastbone, and then forcing the rib cage apart. You stop the heart and cool it. You hook up the patient to a heart-lung machine, which takes charge of breathing and blood routing, but also increases the risk of stroke and kidney failure. When you're done, you wire the breastbone back together and restart the heart. It's no surprise that after such trauma, the patient may need up to a year to recover fully.

Anyone who has witnessed the effects first-hand on a loved one knows that conventional cardiac surgery delivers a ferocious blow to the body.

Surgeons have been performing open-chest surgery in this way for the last thirty years. To change the status quo by performing closed-chest surgery on a beating heart was our Mount Everest. On that September day, we found ourselves finally on the top of the mountain—and on top of the world.

Despite meticulous planning and mapping, however, the climb had not been easy, or fast. In fact, my part in the trek had started years earlier, as a medical student at the University of Ottawa in the 1980s. There, I studied with Dr. Wilbert Keon, founder of the Ottawa Heart Institute. I admired not only his surgical technique, but also how he got things done. He had vision, he knew where he wanted to go, and he wasn't afraid to break new ground. I was there taking notes on May 1, 1986, when Dr. Keon performed the first artificial heart implant in Canada on Noella Leclair, a forty-one-year-old woman from Orleans, Ontario. The operation saved her life. I did not know it at the time, but it had changed mine, too.

By the mid-1990s, I was a transplant and artificial heart specialist with my career path clearly marked out, or so I thought. Change was in the wind, though, and in 1996 I went to England to

To change the status quo by performing closed-chest surgery on a beating heart was our Mount Everest. On that September day, we found ourselves finally on the top of the mountain—and on top of the world.

watch the first beating-heart surgeries and came home inspired. Encouraged by Dr. Keon, I would soon perform one of the first beating-heart surgeries in Canada, which set me on a new—and even rockier—road leading into new territory.

That new territory was closed-chest robotic heart surgery. To master the new surgical territory, three skills are crucial. The first is expertise in operating on a beating heart, which is much more difficult than operating on a still heart. The second involves endoscopic skills—the ability to operate while looking at a monitor's screen and seeing an image sent from a miniature camera inside the chest. The third, minimally invasive surgery, is an operating technique that involves manipulating long-handled instruments through small incisions in the chest—an alternative to the more radical opening of the chest.

Not all the developments were focused on the operating room, however. While surgeons were developing these techniques, surgical robotics had been hurrying along its own path. Initial research during the eighties was funded primarily by the National Aeronautics and Space Administration (NASA) in the United States. People at NASA were already thinking ahead to the day when astronauts would be in outer space for long periods of time without access to medical care. They were planning for the worst. For instance, on Earth, an appendicitis attack would mean an emergency trip to the hospital. In space, it could be disastrous. NASA thought that if it could develop the right type of telerobotics, then, in a medical emergency, a surgeon on Earth could perform lifesaving surgery on a patient in outer space, through a technician and a portable surgical unit.

Later, when the scientists who worked with NASA spun off the technology to private companies, the race was on to see who could develop a surgical robot for civilian use. While the technological development for surgical robots occurred exclusively in California, clinical development in the mid-nineties took place in Europe, particularly in Munich and Leipzig. With generous funding in hand, these German centres were able to put robots in their operating rooms almost as soon as the manufacturers could ship them.

In 1996, inspired by the developments in Germany, I joined the London Health Sciences Centre in London, Ontario, set an agenda, and acquired a partner. Dr. Reiza Rayman shares my desire to explore new territory, and to develop new, robotics-based procedures that will improve the quality of life for so many patients. With his keen business sense and organizing abilities, he has been critical to the success of our program. Our first priority was to define our goal, which was to perform a beating-heart, closed-chest coronary artery bypass. We set down a list of milestones and target dates, and sketched a detailed roadmap. We were on our way to Everest.

Throughout 1996 and 1997, we worked on minimally invasive surgery (MIS) and endoscopic skills. The main problem here was that the surgeon only has two hands. To perform surgery in a closed chest, you need another person to hold the camera while you operate. This means somebody else controls what you see—not the most intuitive way to work.

We were ready to take the next step. We were one of the earliest North American centres to adopt surgical robotics (the first was Penn State in the U.S.). As soon as Health Canada approved the technology for clinical use in a research setting, we took it and leapfrogged the German groups, who had been doing closed-chest surgery on still hearts. In September 1998, the London Health Sciences Centre team used the voice-activated Automated Endoscopic System for Optimal Positioning (AESOP) camera system to help harvest the chest artery that we would use as our heart bypass conduit. Following an incision between the ribs and, under direct vision, we sewed the bypass graft onto the beating heart. It was the first such surgery ever performed in Canada.

The robotic arm we used in that operation was from the California company Computer Motion, and funded by our foundation at the London Health Sciences Centre. With the support and funding in place, we were ready. By this time, a newer computer-enhanced surgical telerobotic prototype (the Zeus Robotic Surgical System)

had been developed by Computer Motion. Substantially more complex and sophisticated, Zeus was made up of three major components: a surgeon interface device consisting of a video monitor and two instrument handles, a computer controller, and two robotic arms. The functions were also broken down into components.

Here's how Zeus works. The surgeon manipulates the instrument handles, which are identical to traditional microsurgical instruments. The computer controller digitizes the surgeon's motions, and relays the information in real time to the robotic arms. These arms are attached to the operating room table, and precisely control the instrument tips that are within the surgical field. A third, voice-controlled robotic arm is used to hold and move the endoscope. With the patient under general anesthesia, you make three pencil-sized holes in the chest. Through these ports, you insert Zeus's three stick-like arms into the chest cavity. One arm holds the voice-activated endoscopic camera while another holds a grasper. The third ends in an ultrasonic scalpel that cuts by vibrating at 55,000 hertz and simultaneously seals over small blood vessels. Not one of the instruments is more than five millimetres in diameter.

The surgeon, wearing a headset and looking more like an airline pilot than a doctor, sits almost two metres away from the patient, and concentrates on a monitor that shows a full-colour, three-dimensional image of the patient's beating heart. The camera moves, obedient to the doctor's voice commands. "In." "Left." "Forward." Tweezer-like instruments, hooked up to a computer controller, sit in the surgeon's hands. They sense every movement and send a digitized signal to the robotic arms. These translate the signal into precise micro-movements inside the patient's body, where the tiny instruments harvest the thoracic artery and sew it to the heart with sutures twice the thickness of a human hair.

Despite this vision of the surgeon presiding in solitary splendour at the console, our work is very much a team effort. I could not have made a move without my surgical colleagues, anesthetists, and nursing staff. Also, without the faith and support of our generous benefactors and our hospital administration, none of it would have happened.

So here we are. We've reached the top of our Everest and the view is breathtaking. That doesn't mean our trek is over, though. Other challenging peaks rise against the horizon.

So here we are. We've reached the top of our Everest and the view is breathtaking. That doesn't mean our trek is over, though. Other challenging peaks rise against the horizon—some near, some much more distant. Which are the peaks to watch? At the National Centre for Minimally Invasive Robotic Surgery (NCMIRS) at the London Health Sciences Centre, watch for the first in-utero fetal surgery. Also watch for the use of robotics in heart surgery for children, and in the surgical treatment of lung cancer and atrial fibrillation (uncoordinated contractions of the heart). The application of robotics to new areas of surgery will ultimately enable us to help our patients in ways we have just begun to dream of.

In time, I believe we will eventually see robots doing heart valve repair as well as multiple bypass surgery—a technique still confined to the lab because of the difficulties of working on the back of the heart. I am confident that all these procedures will, one day, be performed by a surgeon sitting at a console half a world away from the patient. We took the first real step in that direction in March 2001, when our London team performed the world's first series of telementored operations, with surgeons ten kilometres apart and linked by a camera and a teleprompter. The technology is available, it just has to be refined.

Whenever computers in medicine make headlines, we hear talk of machines replacing human doctors. The truth is that robots do not replace surgeons, they enhance a surgeon's capability. Robotics allows us to perform delicate, reconstructive surgeries that could not be done before. Using tiny instruments and high magnification, surgeons can now sew smaller vessels and do finer work than could ever be possible with the human hand. This is a genuine milestone.

The big winner in all this is the patient. An increasing number of patients will now be eligible for more complex operations with less trauma, fewer post-operative complications, better outcomes,

and swifter recovery. Closed-chest patients are now getting out of hospital in two or three days, compared to six or seven days for open-chest surgery. Also, they are going back to work in days or weeks instead of months. With shorter hospital stays and quicker recoveries, the benefit for us all in reduced health care costs will be enormous. For this to happen, however, we will need to train more surgeons and invest in more equipment and machines.

Despite this vision of the surgeon presiding in solitary splendour at the console, our work is very much a team effort.

What is Canada's place in this field? By merit of what we have accomplished over the last few years, we are recognized among the world's leaders, but we can't rest on our laurels. We are extending our research capacity and are busy training a new generation of researchers and surgeons. If Canada continues to support innovation, and to invest in infrastructure, then we will keep our place at the top of the mountain.

Further Reading

"Remote Control Surgery," Gretel H. Schueller, *Science World Magazine*, Vol. 57, No. 5 (2000), pp. 18–21.

Dr. Martha Salcudean was born in the wrong place at the wrong time—in Eastern Europe during the terrible pre-war years when Adolf Hitler appeared unstoppable. The ravages he so extensively inflicted on the world took their toll on Salcudean and her family. After the war, another era of despotism took hold of Eastern Europe. During this time, Dr. Salcudean studied in Romania, earning her doctoral degree, and worked as a senior researcher at the National Research Institute for Metallurgy in Bucharest. After getting out of Romania, Dr. Salcudean and her family reached Canada in 1976. Today, she says she still feels immense gratitude for such things as true democracy, free newspapers, respect for the individual, and being in control of her life.

Dr. Salcudean started her academic career at the University of Ottawa, as an assistant professor in 1976 and then as a full professor in 1980. She established a strong research program, working in the area of fluid flow and heat transfer, and mathematical modelling of processes, and she collaborated extensively with industry. For her accomplishments, she received an honorary degree from the University of Ottawa in 1992.

In 1985, she was appointed head of the Department of Mechanical Engineering at the University of British Columbia, and in 1993 became Associate Vice-President, Research. During those years, Dr. Salcudean led a large research team, establishing a world-class centre of process modelling in the pulp and paper area. In 1996, she was awarded the Weyerhaeuser Chair in Computational Fluid Dynamics, which she currently holds.

In addition, Dr. Salcudean was awarded the 1991 Science and Engineering Gold Medal, in the Applied Science and Engineering category, from the Science Council of British Columbia. She is a Fellow of both the Royal Society of Canada and the Canadian Academy of Engineering. In addition, she was awarded the Commemorative Medal for the 125th Anniversary of Canadian Confederation; the Engineering Institute of Canada's 1995 Julian C. Smith Medal for "achievement in the development of Canada"; the Association of Professional Engineers and Geoscientists of the Province of British Columbia's Meritorious Achievement Award for 1996; the National Izaak Walton Killam Memorial Prize in Engineering in 1998; and the Order of British Columbia in 1998. In 2001, she received an honorary degree in science from the University of British Columbia.

Engineering: Applied Ingenuity

Martha Salcudean

Engineering is an old profession that has always played an important role in making life easier, richer, and safer. From the roads and structures of the Roman Empire, to the 1969 landing on the moon by Apollo 11, we have clear evidence of the significant contributions that engineers have made here on this planet—and beyond. These important technical achievements are a testament not only to their mastery of mechanical and physical laws, but also to their ingenuity, imagination, and creativity.

The public tends to have an image of the engineer as a rigorous analyst, a person who can express all the known laws of physics in order to build artifacts that we can use to make life easier, safer, and more enjoyable. This image is far from realistic. It is amazing how many incredible inventions have been introduced based on an idea and a dream, but without a complete understanding of all the physical phenomena on which they are based. For instance, aircraft preceded a full understanding of aerodynamics.

Despite the publicity given to accidents, it is actually quite amazing how well the things built by engineers have worked, and continue to do so.

Papermaking machinery was used well before there was a full understanding of the laws governing the flow of pulp. Ingenuity, experience, learning from mistakes, and a mixture of a few big steps and many incremental improvements have produced a great many devices and processes that serve mankind well.

Engineers have always tried to understand the laws of physics and apply them to what they were building or making. However, the complexity of most engineering feats was far greater than what could actually be calculated. To overcome this, engineers tried to simplify problems, often to unrealistically simple forms and surroundings. Then they applied caution in the form of "safety factors," paying a cost in performance as well as in price. Sometimes, of course, engineers got it wrong and a big price was paid—in human life, in inconvenience, in money. But most other times they got it right. Despite the publicity given to accidents, it is actually quite amazing how well the things built by engineers have worked, and continue to do so.

It is important to remember, for example, that not so long ago there was no way to accurately define the temperature of a fairly simple shape, such as a cube heated in a furnace. The calculations required to do the description were understood, but were just too complicated to be completed accurately within any reasonable time frame. To compensate for this inability to make precise calculations in such circumstances, engineers drew on experience, conducted experiments, and transmitted the ever-improving knowledge from one generation of practitioners to the next. Lots of hard work helped to establish all kinds of half-theoretical, simplified solutions, which were adjusted with some empirical factors arrived at through trial and error.

Now that we find ourselves in the computer age, we have the possibility of making very complex calculations with such accuracy that the way engineers design products and processes is being revolutionized. Better and safer products can now be made in far less time. John von Neumann, a genius mathematician and early computer scientist born in Budapest, Hungary, in 1903, foresaw the

huge impact computing would have on our engineering knowledge. Before he passed away in 1957 in the United States, he predicted a time when we would design our large aircraft using computers, and see the results of changes we made in real time, even as we made them. His predictions have, of course, come true.

When I worked as a researcher in Romania at the end of the 1960s, I was very attracted to the possibilities offered by the computer. One day, as I observed a steel-making process, I watched as huge steel blocks, some two-and-one-half metres tall and weighing as much as twenty-five tons, were endlessly reheated in furnaces to make sure they would be hot and pliable enough to roll into much flatter and thinner slabs. It was hard not to be impressed by the rolls pressing the incandescent red metal that obediently took the desired shape. Far too often, however, the steel was not hot enough inside. As a result, the pressing rolls could not compete with the strength of the metal, and the whole process would come to a grinding halt. A couple of colleagues and I thought that the process would work much better if the metal was hotter on the inside than it was on the outside. The trouble with this idea was that we had to come up with a way of achieving this, because using a furnace would always cause a cooler temperature on the inside.

After some thought, we came to a conclusion. The idea was to not lose the heat inside the block to start with. We figured that if we could quickly transport the block into the furnace, before it was completely solid on the inside, it would be heated from both inside and outside. The block would heat up quickly and would roll like a dream. Moreover, the process would require much less fuel. We were proud to have figured out this solution, but there was still one problem with our proposal. If by any chance a block that had liquid inside passed between the rolls, the whole plant would be brought to a spectacular and catastrophic halt. How could we figure out whether the inside was fluid or solid? We couldn't measure the inside temperature because the instruments did not yet exist.

Faced with a new challenge, we then decided that I would use the only IBM 360 available in Bucharest at the time to solve the problem mathematically. Remember that earlier we could not calculate the temperature in a simple solid cube, and now we had to calculate the temperature of an ingot that was half solid and half liquid and undergoing solidification. I developed the computer model, ran the program, decided on the proper heating time, and took the result to the steel makers. I still remember standing on the bridge of the facility watching the ingot going through the rolls, with my heart rate close to its upper limit. After all, in a communist country, the "reward" for inflicting significant damage to a steel plant was much more far-reaching than a few bruises to one's ego.

Thankfully, it worked, and it gave me great satisfaction. I found myself happy, and proud of my decision to become a mechanical engineer. Ever since, I have more or less stuck to developing models for all kinds of interesting processes, and products for all kinds of industries. It's quite amazing how many different processes there are to which one can apply these methods. But, of course, one needs to thoroughly understand the processes being modelled. That takes a substantial amount of work.

In a communist country, the "reward" for inflicting significant damage to a steel plant was much more far-reaching than a few bruises to one's ego.

After immigrating to Canada, I had more opportunities to work on process modelling, which I did for the steel industry in collaboration with Professor Rod Guthrie from McGill University. My move to the University of British Columbia in the mid-1980s provided other, related opportunities. I have been fortunate that the Natural Sciences and Engineering Research Council has generously supported my work. In collaboration with my colleague, Dr. Ian Gartshore, I have worked for several years on a project with Pratt and Whitney Canada that involves turbines for aircraft.

Of course, it is paramount for aircraft engines to be fuel-efficient, as they carry people from one part of the world to another. Fuel efficiency is important not only because it affects the price of air transportation, but also because the less fuel an engine needs, the less pollution it will cause. Turbines in aircraft engines are more efficient if they can work with gas entering the turbine

Fuel efficiency is important not only because it affects the price of air transportation, but also because the less fuel an engine needs, the less pollution it will cause.

blades at higher temperatures. However, there are clear limits to increasing these temperatures. For example, the blade—which is like a small aircraft wing—can deform under high temperatures, which creates a serious problem. This problem can be solved in two different ways. First, we can use a very expensive, exotic, heat-resistant material to build the blade. Second, we can try to blow cold air right over the blade surface to protect it, like a curtain between the blade and the hot air. My colleague and I have spent quite some time determining how best to inject this air so that it stays on the surface of the blade, rather than rushing into the hot stream. We have made computer models of the blade and included the internal and external flows, finally coming up with an optimum arrangement.

Another area that I have had the opportunity to work in is the pulp and paper industry in British Columbia. A very important industry for the province, it was even more important at the time of my move to B.C. in the mid-1980s. When I moved there to work on process modelling for pulp and paper production, I thought it would be mutually helpful to find partners in the industry. The engineering problems in this industry are varied. How can we make better-quality paper that is more satisfactory for today's beautiful colour prints? How can we eliminate the ugly spots on paper? How can we make paper mills more efficient, so they run with minimum impact on the environment and with the least unpleasant smell?

We concentrated first on a piece of equipment called the "recovery boiler." A Canadian invention, the recovery boiler is very large, with a base size of about ten by ten metres, and about thirty metres high. The fuel burned inside the boiler furnace is called "black liquor." This intriguing name was given to it because of a chemical that comes from the process of making wood chips into paper, and which becomes black by dissolving the lignin in wood chips. Because it contains precious fuel and chemicals that can be reused, black liquor cannot be thrown away. It would also cause significant environmental damage if it were simply dumped into a lake or river.

The black liquor is burned in the recovery boiler in order to use the fuel it contains while recovering chemicals that can be re-used. This must be done in a way that recovers the chemicals while, at the same time, producing large amounts of hot water and steam to generate power in the pulp and paper mill—and all this must be accomplished with maximum efficiency, minimum pollution, and the greatest possible safety for operators and the public. My colleague and I were able to develop the necessary models, and have applied them successfully to improve many recovery boilers. As a result, we have increased knowledge and expertise in this area for both equipment manufacturers and users. Dr. Peter Gorog, from Weyerhaeuser Company, has played an essential role by helping with the development and funding of our research program, and even more critically by promoting and championing extensive industrial applications of our results. Without his belief in technology and his unwavering support, the impact of this research would have been very much reduced.

What do all these processes have in common? How can we approach such a variety of applications using similar methods? The same laws of physics that teach us how liquids or gases move, and how heat and chemical species are carried, govern all these problems. The laws have complex mathematical expressions, but the calculations can all be solved using computers. We have the ability to solve equations for hundreds of thousands of points, all within hours or a few days. Even better, because of the increasing speed and memory of today's relatively inexpensive computers, we can now do in a few hours what, just a few years ago, took days with very expensive machines.

We are looking at developing this technology further within both our research group and Process Simulations Limited, a UBC spin-off company started by Dr. Gartshore, Dr. Zia Abdullah, and myself. We have been joined at Process Simulations by talented collaborators David Stropky, Eric Bibeau, and Jerry Yuen. We are developing such products as virtual cameras that allow the operator of a recovery boiler to see the combustion taking place inside the furnace. As well, we are developing simulators that train recov-

ery boiler operators effectively and allow the full range of the boiler's operations to be simulated, including accident scenarios that could never be demonstrated on real units.

We believe that within the foreseeable future—as background process knowledge improves, mathematical techniques evolve, and computers become faster—all processes will be simulated, and a simulator will be a normal tool used by designers and operators alike. The opportunity created to develop new products and processes, and to improve existing ones, is more extraordinary than ever. It will continue to challenge engineers to put their ingenuity, imagination, and creativity to work.

Because of the increasing speed and memory of today's relatively inexpensive computers, we can now do in a few hours what, just a few years ago, took days with very expensive machines.

Further Reading

From Steam to Space: Contributions of Mechanical Engineering to Canadian Development, edited by Andrew H. Wilson (Ottawa: Canadian Society for Mechanical Engineering, 1996).

"Modelling of Industrial Processes Using Computational Fluid Dynamics," Martha Salcudean, *Canadian Metallurgical Quarterly*, Vol. 37, No. 3–4 (1998), pp. 251–63.

"Computational Fluid Flow and Heat Transfer—An Engineering Tool," Martha Salcudean, *Transactions CSME*, Vol. 15, No. 2 (1991), pp. 125–35.

A maritime historian and an author of numerous books on the shipping industry in Atlantic Canada, *Dr. Eric W. Sager* is a professor of history and chair of the Department of History at the University of Victoria. From 1996 to 2001, he expanded his scope of interest and specialization with a stint as director of the Canadian Families Project, a major collaborative initiative funded by the Social Sciences and Humanities Research Council.

A member of the Atlantic Canada Shipping Project at Memorial University of Newfoundland from 1976 to 1982, he is the author of *Seafaring Labour: The Merchant Marine of Atlantic Canada,* 1820–1914 (McGill-Queen's University Press, 1989); co-author, with Gerald Panting, of *Maritime Capital: The Shipping Industry in Atlantic Canada, 1820–1914* (McGill-Queen's University Press, 1990); and author of *Ships and Memories: Merchant Seafarers in Canada's Age of Steam* (UBC Press, 1993). In addition, he is co-author, with Peter Baskerville, of *Unwilling Idlers: The Urban Unemployed and Their Families in Late Victorian Canada* (University of Toronto Press, 1998).

Dr. Sager has published many articles on the history of shipping, the history of labour, and the history of families. He received a B.A. in 1966 and a Ph.D. in 1975, both from the University of British Columbia. He teaches courses on Canadian history, labour history, family history, and the use of computers in history.

Canada's Families: Why History Matters

Eric W. Sager

Fifteen years ago, if anyone predicted that I would become the director of a project on the history of the family, I would have said they were joking. At the time, I was writing a book on the history of Canadian sailing ships. I was a *maritime* historian. I had a great research subject, and plenty to write and talk about. Yet within a few years, I was planning a collaborative research project on the history of the family. What accounts for the big switch? Why would a historian of ships and sailors become fascinated by families and their history?

The quick answer may be that I am an intellectual gadfly, skipping from one subject to the next, but the better, more thoughtful answer has to do with history itself, and why historians do what they do. History is a dialogue between past and present. It is always an attempt to make sense of the past, in and for present times. Although history is useful—in the narrow sense that it offers clear lessons about how we should behave today or how we can avoid

repeating past mistakes—its value is much deeper. History is an attempt to shape and inform our collective memories, our shared understanding of human experience. Without history, we would suffer a profound amnesia—the loss of all informed memory of human experience in time and space.

My journey into family history was part of my ongoing dialogue with the past and the present. In that conversation, the present always has its say. Today's concerns about families influence historians and guide our curiosity. How often do we read today that there is a "crisis" in the family? In the words of a famous American sociologist, "Some have cited facts such as the high rates of divorce, and changes in the older sexual morality ... as evidence of a trend towards disorganization in an absolute sense."

We read that modern living arrangements cater to single people or couples, leaving no room for the three-generation "extended" family. A Canadian political economist wrote that "The home has passed, or at least is passing, out of existence. In place of it is the apartment—an incomplete thing, a mere part of something, where children are an intrusion, where hospitality is done through a caterer, and where Christmas is only the twenty-fifth of December."

We also read about the selfish individualism of a "me" generation. As a Protestant minister wrote, "We may expect to see further disintegration until the family shall disappear.... In all things civil and sacred the tendency of the age is towards individualism ... its plausible aphorisms and popular usages silently undermining the divine institution of the family."

But are we in the present or the past? The three passages I have quoted, although they echo present fears, come from the past, not the present. My first quotation is from the American sociologist Talcott Parsons, writing in 1955; my second is from the Canadian political economist and humorist Stephen Leacock, writing in 1915; and the third is from a Protestant minister in Ontario, writing in 1878.

The point is not that the past and present are the same, or that history repeats itself. The point is that history lets us see what was previously hidden. The supposed "crisis" in the family has been with us for a long time. In every generation, for more than a century, a vocal minority has predicted the death of the family. So far, every rumour of its death has been premature. As a result, history compels us to ask not whether the family is disappearing, but rather why different generations in North America have become so fearful. What changes in families and households really lie behind the rhetoric, the nostalgia, and the conflicting values surrounding the complex social unit that we call the "family"?

My engagement with family history comes from the intersection of present-day public debates with my intellectual interests, each guiding the other. After finishing my studies of sailing ships, I embarked on a study of unemployed people in late-nineteenth-century Canada. I did this after my colleague, Peter Baskerville, and I noticed that the 1891 census asked adult Canadians whether or not they were unemployed. Our curiosity was piqued. We wanted to identify who the unemployed were, and how they coped in the days before social welfare and unemployment insurance. The further we searched, the more we realized that we first had to understand the family context of the unemployed. Most previous studies of living standards had been limited by their focus on individuals, but all workers had families or households. Most of them, even the single men in logging and mining camps, shared living spaces or pooled resources with others. So I began to read the works of historians and others who wrote about the family economy.

What changes in families and households really lie behind the rhetoric, the nostalgia, and the conflicting values surrounding the complex social unit that we call the "family"?

I was also aware of the increasing intensity of political debates about family in the United States and Canada. American politicians were cutting welfare support for single mothers in the name of "family values." Canadian writers sounded alarm bells as divorce rates rose and the poverty of single parents became a public issue. The United Nations declared 1994 to be the International Year of the Family, but in Canada, major groups interested in family matters could not even agree on a definition of "family." It was at that time that the subject took a personal turn. I found myself contributing to certain trends when my marriage ended and I joined the ranks of single parents. I began to read the literature on children of divorced parents, this time for more than pure

I found myself contributing to certain trends when my marriage ended and I joined the ranks of single parents. I began to read the literature on children of divorced parents, this time for more than pure academic interest.

academic interest. The personal, the political, and the intellectual had converged.

In the early 1990s, the original returns for the 1901 census of Canada became available to researchers. For the first time, we could see the answers that Canadians gave to enumerators and enumerators wrote on their census schedules. Suddenly, we had a source of data about the people in my great-grandparents' generation that was unprecedented in its richness. For the 5.3 million people on the census list, the enumerators had tried to record name, age, sex, birthplace, nationality, racial origin, occupation, religion, earnings, literacy, mother tongue, property owned or rented, and much more. This census was even richer than those conducted in Britain and the United States during the same time. Here was a unique opportunity to view individual Canadians, together with their families and households.

Was family a safe haven for the unemployed and the poor? Did some family members make up for the low wages or unemployment of other family members? A century ago, the answer in six Canadian cities was a resounding no. When all earnings were pooled, at least one in seven working-class families lived below a minimum-survival poverty line. The informal economy was much larger than it is today. Scrounging, bartering, peddling, taking in laundry, renting rooms to boarders, and gathering wood or cinders for fuel were the daily activities of large numbers of women and children. It's little wonder that death rates were high. In the working-class districts of our cities, one in three or four babies died before having a first birthday. Although the family was the first line of defence against poverty and unemployment, it was, by itself, a very weak defence.

While writing our book about the unemployed, *Unwilling Idlers*, Peter Baskerville and I met others who were using historical census data. We realized that we had barely scratched the surface of the 1901 census. We had looked at the unemployed in six cities, but could we observe *all* Canadian families, both rural and urban? Of course we could, we concluded. After all, American scholars had created national samples using their census returns from 1850 to the present. Inspired by the example of Steven Ruggles and his team at the University of Minnesota, we assembled a team of eleven scholars (historians, sociologists, geographers, and demographers), and applied to the Social Sciences and Humanities Research Council for a grant to computerize a national sample of the 1901 census. Our first application failed, but in 1996, our second application succeeded.

The early results of our work marked a sudden leap forward in our knowledge of Canadian people and Canadian families. There was no single "traditional" family, we discovered. Although a majority of households a century ago consisted of two parents and their biological offspring, over a third of households were of other types. Most people lived in a wide variety of household types over the course of their lives. The pattern varied among regions and among ethnic groups. In British Columbia, for example, a third of households consisted of men only, and a third of the population did not live with any family members (relatives by kin or marriage). Today's "non-traditional" households—the households of people of the same sex, single-parent households, cooperative households of unrelated people—all have their own traditions. When we use Statistics Canada's current definition of "family" for census purposes, we find that there were as many single-parent families in 1901 as there were in 1996.

These results help us to dispel myths and misconceptions based on nostalgia and individual memory. Change is a constant, but change is not the same as crisis. The family form is variable and flexible, and it remains intact. We know family by what it does, not by any single traditional form. For a moment, think about one of English Canada's most famous families from a century ago. A fictional family, to be sure, but a believable one. It consists of an elderly spinster, her brother, and a non-kin child—Anne of Green Gables. By Statistics Canada's definition of "family" on its census, this small group is not a family. Yet we all know it to be a family, not by its form but by what these people did with and for each other.

Today, I am part of a team of scholars (the Canadian Century Research Infrastructure Project) that seeks to study Canadians and Canadian families across the whole of the twentieth century, using twentieth-century censuses. We have a powerful model to follow. In the United States, scholars have integrated all decade census samples into one database: the Integrated Public Use Microdata Series (IPUMS). And Steven Ruggles and his Minnesota team are taking another big step forward in their work to integrate census databases from more than twenty countries.

Will Canadians be part of this massive international effort? The answer depends on whether we fund research of national importance, and how we allocate our research funds. There is a great deal at stake in terms of national self-understanding. For example, was the social welfare system responsible for the dramatic rise of single-parent families in the twentieth century? Americans can answer the question with a precision that Canadians can only dream of. That's because American scholars can use individual-level census microdata to connect changes in welfare payments to changes in family structure over decades. "We must conclude," says Steven Ruggles, speaking about the United States, "that the dramatic increases in divorce, separation, and illegitimacy since the Great Depression cannot be attributed to the growth of welfare." At present no such definitive answer is possible for Canada.

When we use Statistics Canada's current definition of "family" ... we find that there were as many single-parent families in 1901 as there were in 1996.

What happened to immigrants and their families after they arrived in Canada? Did they find stable employment, or did they, in significant numbers, endure poverty and welfare dependence? Only an integrated public use census series for Canada will give us a complete answer. Other questions still linger as well. Where and why did infant mortality rates decline? What was the real rate of marital separation and single parenting prior to the Divorce Act of 1968? In what social, economic, and family circumstances does women's participation in the labour force rise? What are the circumstances that affect changes in housing conditions and housing stock?

These questions, and many others, will be answered in the coming years if Canadian scholars can join the great international collaborations now being created. We may be able to answer these questions in an entirely new way by comparing the Canadian experience to those of many other countries. This is the journey of self-understanding that historians challenge us to take.

Further Reading

Unwilling Idlers: The Urban Unemployed and Their Families in Late Victorian Canada, Peter Baskerville and Eric W. Sager (Toronto: University of Toronto Press, 1998).

The Infinite Bonds of Family: Domesticity in Canada, 1850–1940, Cynthia R. Commacchio (Toronto: University of Toronto Press, 1999).

Profiling Canada's Families II, The Vanier Institute of the Family (Ottawa: Vanier Institute, 2000).

The Way We Never Were: American Families and the Nostalgia Trap, Stephanie Coontz (New York: Basic Books, 1992).

The Way We Really Are: Coming to Terms with America's Changing Families, Stephanie Coontz (New York: Basic Books, 1997).

For the Canadian Families Project, see the special issues of the following journals:

Historical Methods: A Journal of Quantitative and Interdisciplinary History (Fall 2000).

The History of the Family: An International Quarterly (December 1999).

Journal of Family History (April 2001).

Dr. David Coleman is a professor and chairman of the Department of Geodesy and Geomatics Engineering at the University of New Brunswick.

Before obtaining his Ph.D. in 1994 from the University of Tasmania, Australia, he spent fifteen years in the Canadian geomatics industry—as a project surveyor and engineer with Marshall Macklin Monaghan, a senior manager with Northway-Gestalt Corporation, and a partner in The Cabot Group, a land information management consulting firm. After working in a number of locations across Canada, Dr. Coleman and his family spent two years in Australia while he obtained his Ph.D. He has authored numerous papers on institutional and market aspects associated with land information policy, geomatics operations management, GIS implementation and assessment, and emerging national and global spatial data infrastructures.

Dr. Coleman is currently president of the Canadian Institute of Geomatics. He has also been a member of the U.S. National Research Council's Mapping Sciences Committee, the Defence Science Advisory Board, the Champlain Institute Board of Directors, and the Research Management Committee of the GEOIDE Network of Centres of Excellence. In addition to his teaching and research at the University of New Brunswick, he is a consultant to clients in Canada, the United Kingdom, and Latin America.

Moving Maps and Geographic Information Systems onto the Internet

David J. Coleman

First the earth was flat. Then it was round.

The discovery was exciting enough to catch the attention of most people on the planet. Today, centuries after that earth-changing revelation, it might appear to many that the most exciting times of exploration and cartography are long behind us. They're wrong.

In fact, it is likely that humanity has never seen as exciting and productive a period of map production and mapping technology development as in these past forty years. This is especially true here in Canada. Geographic Information Systems (GIS), which Canadians have had a big hand in inventing, have helped put maps and related information to work in everything from municipal engineering and forest management, to health care planning and market research. As new mapping and positioning technologies move onto the World Wide Web and into wireless, location-enabled appliances, we are beginning to see a fundamental change in the

way people apply geography in their everyday lives. To those of us with a love of maps and geography, we find ourselves in the middle of fascinating and uncharted territory.

As new mapping and positioning technologies move onto the World Wide Web and into wireless, location-enabled appliances, we are beginning to see a fundamental change in the way people apply geography in their everyday lives. To those of us with a love of maps and geography, we find ourselves in the middle of fascinating and uncharted territory.

In fact, I have long been fascinated by maps and mapping. As a seven-year-old "official navigator" on family road trips; as a student listening to his teacher explain how the moon was mapped; and as a young engineer working with a Canadian aerial mapping firm—at every step of the way, I have tried to soak up the history and experiences of mapping, wherever they were found.

Today, at the University of New Brunswick, my fascination with maps continues. As a university professor in geomatics engineering, I help design the systems and practices people use to collect, analyse, visualise, and use all kinds of geographic information. I am fortunate enough to have turned my passion into a rewarding career; and now, with developments such as GIS, and mapping and positioning technologies, the future (and my chosen vocation) promise more excitement than ever.

Before we take a closer look at the exciting road that's carrying us into the future, however, we should take a short trip to our past, so that we may better understand the history of mapping in our vast and geographically varied country.

Canadians have long held an interest in collecting, organizing, and using land-related information in the establishment and development of their country. From early mapping efforts in the 1600s and 1700s by the French and English for defense and settlement, to the systematic inventories of natural resources, to the detailed urban and property mapping programs of the post-

war era, Canadians have entrusted their governments with responsibilities for detailed surveys, mapping, and land administration activities.

Modern-day Canadians have also figured prominently in the development of digital mapping and GIS. Developed in the sixties and seventies, early versions of these systems gave their users the opportunity to relate maps to a common referencing system. The maps could then be linked within a single database to descriptive information concerning physical topography, sub-surface soils and geology, forestry, wildlife, and even property ownership. No small task.

Then, beginning in the mid-eighties, higher-performance workstations—connected through local and wide-area networks—became a viable alternative to host-based and stand-alone PC-based configurations. By the mid-nineties, broadband was already beginning to make network-based data delivery and software support practical for everyday users in the GIS community.

From early mapping efforts in the 1600s and 1700s by the French and English for defense and settlement, to the systematic inventories of natural resources, to the detailed urban and property mapping programs of the post-war era, Canadians have entrusted their governments with responsibilities for detailed surveys, mapping, and land administration activities.

It has, however, been the ease of use and the multi-media capabilities of the World Wide Web, and the resulting increases in usage of mapping technologies, that have attracted most of the mass-market attention since 1994. In combination with software packages like Netscape and Internet Explorer, the Web quickly became a mainstream tool for online ordering and distribution of land-related information, custom map creation and display, and Web-based GIS.

Today, most GIS users in developed countries use the Internet to access a wide variety of public-domain and commercial spatial data sources related to transportation, navigation, property, natural resources, and market demographics. Canada is no exception. With the introduction of the Real Property Information Internet Service

(RPIIS), New Brunswick became the first Canadian province, and one of the first jurisdictions anywhere in the world, to offer complete and integrated, Web-based access to its digital property mapping, ownership, and assessment information holdings.

The path leading to New Brunswick's system began in early 1994, when research staff at Universal Systems Limited (USL) in Fredericton, and the Department of Geodesy and Geomatics Engineering at the University of New Brunswick, began joint development of wide-area network–based enhancements to USL's existing CARIS GIS software. As it was originally conceived, the new software would allow people to use standard indexing and searching tools to identify, locate, select, and download an organization's map products across a dial-up network. That was the first step.

Rapid advances in wireless communication and palmtop computing, and the integration of GIS and global positioning system (GPS) technology, have extended the application of traditional GIS databases out into field operations.

Following subsequent research and development, supported by the Canadian Network for Advanced Research and Industrial Excellence (CANARIE) program and by USL itself, the product definition and preliminary software development stages were completed by early 1995. After an extended period of testing, USL commercially released the CARIS Internet Server software in mid-1995. According to authoritative sources, it was the first commercial release of such software anywhere in North America, and it had taken place in New Brunswick.

The software was also of immediate interest to the province of New Brunswick. Given the province's long history of examining alternative approaches to online access to, and distribution and sharing of, the land- and property-related information maintained by different government departments, the software seemed a natural fit. Today, Service New Brunswick (the province's Crown corporation responsible for managing the land administration infrastructure for the province) uses the software to take care of operating the real and personal property registration systems; assessing all land,

buildings, and improvements for the provincial property tax system; maintaining the province's surveying and mapping systems; and providing land and geographic information services to the public.

Judging by customer feedback, the RPIIS was an immediate success. By offering subscribers Internet access to property-related assessment and registry information on a transaction-by-transaction basis, users in both urban and rural communities were given a novel way to obtain the information they needed for property transfers, major engineering projects, and general inquiries. Within three years of its inception, virtually all government departments, and most major developers, utilities, law firms, assessors, and surveyors in the province were registered subscribers.

By 1999, Service New Brunswick had added the "land gazette." This enabled government agencies, departments, and municipalities to serve public notice of time-limited interests and notices for parcels of land by linking information directly to the Parcel Identifier (PID). Through a cooperative arrangement with the New Brunswick Department of Environment, users can get information about land parcels online. For example, they can find out if the parcels have underground petroleum storage tanks, are located in municipal watershed areas, or have been part of former dump sites (see Figure 1 on page 50).

As soon as the new technology was up and running, it was time to take it out the door—literally. Rapid advances in wireless communication and palmtop computing, and the integration of GIS and global positioning system (GPS) technology, have extended the application of traditional GIS databases out into field operations. Using regular telephone lines and, more recently, cellular telephone modems, field users can connect to enterprise-wide networks and get access to information resources that once were only available from within a branch office or headquarters.

Using regular telephone lines and, more recently, cellular telephone modems, field users can connect to enterprise-wide networks and get access to information resources that once were only available from within a branch office or headquarters.

Figure 1 • RPIIS Land Gazette: Identification of Petroleum Storage Tanks

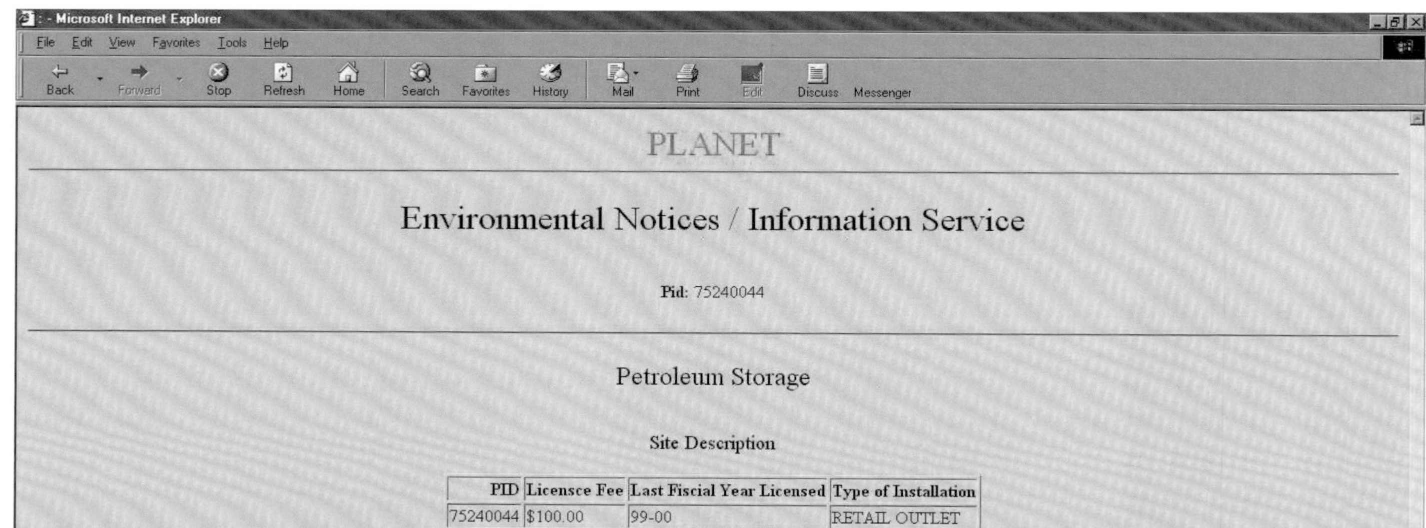

PLANET

Environmental Notices / Information Service

Pid: 75240044

Petroleum Storage

Site Description

PID	Licensce Fee	Last Fiscial Year Licensed	Type of Installation
75240044	$100.00	99-00	RETAIL OUTLET

Tank/s Description

Tank Number	Year Installed	Type	Status	Construction Material	Protection	Capacity (litres)	Product Stored	Pipe Type	Pipe Protection
1	1979	Under Ground	Removed	Steel	Sacrificial Anode	45500		Galvanized Steel	Cathodic
2	1979	Under Ground	Removed	Steel	Sacrificial Anode	45500		Galvanized Steel	Cathodic
3	1979	Under Ground	Removed	Steel	Sacrificial Anode	27000		Galvanized Steel	Cathodic
4	1966	Under Ground	Removed	Steel	Sacrificial Anode	27200		Galvanized Steel	None
5	1966	Under Ground	Removed	Steel	None	17000		Unknown	None
6	1980	Above Ground	Active	Steel	None	910		Copper	None
7	1980	Above Ground	Active	Steel	None	910		Copper	None
8	1979	Under Ground	Removed	Steel	Sacrificial Anode	4546		Unknown	None

For further information you can review the Petroleum Storage Regulation summary page

No matter how familiar the earth we walk on becomes—be it flat, round, or wired—there is always new, uncharted territory to explore.

The transition is now taking place on a number of exciting fronts. New tools and applications are becoming available for working with spatial datasets inside larger, online "data warehouses," with applications in forestry, facilities management, and health care. With the help of the Canada Foundation for Innovation and other partners, the University of New Brunswick is developing an extensive, Web-accessible spatial data warehouse of its digital map holdings. Research funded by CANARIE and, more recently, the Geomatics for Informed Decisions (GEOIDE) Network of Centres of Excellence is developing at least one business-to-business application of networked GIS, applied to digital map production management and quality control. Also, a wide variety of mass-market wireless mobile devices—mobile phones, pagers, dashboard car navigation systems, personal digital assistants, laptop and pocket PCs, and even wristwatches—have been "location enabled" to some degree. This location enabling is facilitated by precise GPS positioning, or by lower-accuracy triangulation through the cellular telephone network. In addition to early services that provide detailed driving instructions to get you from where you are to where you want to be, other applications may be found in retail comparison shopping, advertising, health care, fleet management, and even security services.

Although we now understand "e-commerce" and "e-government" to mean the delivery of certain retail and administrative services across the Web, the term "e-governance" represents a promise of providing citizens with the relevant information, easy-to-use modelling capabilities, and feedback mechanisms they need in order to play a more important role in government decision-making. Research is now underway at the University of New Brunswick, the University of British Columbia, and elsewhere that offers examples of how Web-based, GIS technology will enable citizens to provide prompt feedback and alternative suggestions to local government officials on sensitive urban and regional planning proposals.

Throughout the 1990s, tremendous advances in broadband and wireless communications technology, as well as the explosion in Internet usage and the extension of Internet browsing technology, have dramatically extended the reach and range of mapping and GIS users working in offices, laboratories, the field, or at home. Canadians in research laboratories, government departments, and high-technology firms have played an important role in these advances. They have also demonstrated that, no matter how familiar the earth we walk on becomes—be it flat, round, or wired—there is always new, uncharted territory to explore.

Further Reading

"Internet Access to Real Property Information," B. Arseneau, A. Kearney, S. Quek, and D. Coleman, Proceedings of the GIS '97 Conference, Vancouver, BC, GIS World Inc. (March 1997).

"Information Access and Network Usage in the Emerging Spatial Information Marketplace," David Coleman and John McLaughlin, *Journal of Urban and Regional Information Systems Association*, Vol. 9, No. 1 (1997).

"Location-based Services: Where Wireless Meets GIS," Joe Francica, *Business GeoGraphics* (2000). Available online at: http://www.geoplace.com/bg/2000/1000/1000spf.asp.

GIS On-Line: GIS Information Retrieval, Mapping and the Internet, Brandon Plewe (Delmar Publishers, 1997).

The GIS Portal Web site provides links to a wide variety of GIS-related sites around the world. Available online at: http://www.gisportal.com/.

The Geography Network offers a link to Web-based mapping sites and datasets from across North America and beyond. Available online at: http://www.geographynetwork.com.

The *National Atlas of Canada* is a wonderful resource that can be used to create custom maps online, download lots of useful information, and learn more about Canada's communities. Available online at: http://atlas.gc.ca/.

Bartha Maria Knoppers is a professor at the University of Montreal's law faculty, and a senior researcher (C.R.D.P.) and counsel to the firm of Borden Ladner Gervais. She received a B.A. from McMaster University, an M.A. from the University of Alberta, an L.L.B. and a B.C.L. from McGill University, a D.L.S from Cambridge University, and a Ph.D. from the Sorbonne, in Paris. She was admitted to the bar of Quebec in 1985.

Professor Knoppers is a Forum Fellow of the World Economic Forum (Davos). She is past commissioner of the Royal Commission on New Reproductive Technologies (1991–94), and was named Visiting Heritage Scientist by the Alberta Medical Research Heritage Fund (1993–95), and Fellow of the American Association for the Advancement of Science in 2002.

She is a member of the External Review Board of the Elizabeth Glaser Pediatric AIDS Foundation, past president of the Canadian Bioethics Society, past vice-president of the National Council on Bioethics in Human Research, and co-chaired the Quebec Bar Committee on the Representation of Children (1993–95). From 1995 to 1999, Professor Knoppers was a member of the Management Committee of the Canadian Breast Cancer Research Initiative (CBCRI), the Canadian Pediatric Tumor Bank (Health Canada), the National Expert Advisory Committee on Xenotransplants (Health Canada), and became chair of the Social Issues Committee of the American Society of Human Genetics (1995–98).

Currently the chair of the International Ethics Committee of the Human Genome Project (HUGO), she was a member of the International Bioethics Committee of the United Nations Educational, Scientific, and Cultural Organization (UNESCO), which drafted the *Universal Declaration on the Human Genome and Human Rights* (1993–97). She is also co-founder of the International Institute of Research in Ethics and Biomedicine (IREB) and co-director of the Quebec Network of Applied Genetic Medicine (RMGA). In 1999, she became a member of the Canadian Biotechnology Advisory Committee, and in 2000, she became a member of the board of Genome Canada, as well as the Ethics Committee of the Canadian Institutes of Health Research.

In 1996, Professor Knoppers chaired the Organizing Committee of the first international conference on DNA sampling. In 1997, she was named "Scientist of the Year" by Radio-Canada and by the newspaper *La Presse*, and received the Medal of the Quebec Bar. In 2000, she was awarded a Canada Research Chair in Law and Medicine, and in 2001 received an honorary Doctor of Laws from the University of Waterloo. In May 2002, Ms. Knoppers received an appointment as an Officer of the Order of Canada.

Professor Knoppers was born in the Netherlands and is married with two children.

Biotechnology, Public Policy, and the Unknown

Bartha Maria Knoppers

"What made you interested in studying ethics and human genetics?" is the question I am asked most often. Not having had time to question my personal trajectory, I usually answer by outlining a series of academic events. At other times, I simply say, "Serendipity."

But given this occasion to reflect on my journey to biotechnology, law, ethics, and policy-making, I can honestly answer that it is primarily the excitement of exploring the unknown. I realize, of course, that this constitutes a simplistic answer to a complex question, one that is both classical and contemporary in nature and proportion—are humans just another form of living matter?

As modern biotechnology crosses all living organisms through the science of genomics, a new industry of the life sciences is emerging. Exposure to present and future environmental concerns has helped us realize not only our interdependence with all living matter, but also our co-evolution and co-adaptation. The need to exercise responsible stewardship is self-evident, but can we also shape the future? Even if we could, should we? And how?

In the face of present and future challenges, it was the task of developing an "ecosystem" framework for stewardship that led me to first begin questioning artificial insemination by donor. That was back in the 1970s, when it was actually a covert practice. The route to this area of research was a circuitous one for someone who had spent seven years, at various universities, studying literature. In fact, while I was working on my doctorate (comparing the poetry of the Caribbean with that of Quebec), the political momentum of the period led me to abandon literature for law. I envisioned law as a motor of social change. Little did I know that, in fact, change often precedes the law. Indeed, the speed of technological change requires the ability to think prospectively.

The media cult of the weekly, if not daily, discovery of the wonders of biotechnology creates a certain awe: Awe in the face of an invention, and awe in the face of the disclosure of the mysterious laws of nature. It also creates the hope, though, that such knowledge will alleviate human suffering. Like the tree of good and evil, however, such knowledge is also seen as potentially dangerous. Suggestions that it could lead to discrimination, stigmatisation, eugenics, and even biological terrorism have all been made. As a result, the challenge is to not succumb to facile polarization due to the questions raised in the biotechnology debate. Indeed, the tendency to reduce complex issues to black-and-white positions with simplistic media clips borders on the unethical. Through its polemic, it tends to foster blanket prohibitions, or it depicts the issues in such a way that they appear insurmountable, with the result being political apathy. Debate may be heated, but it is vacuous and sterile. Although they share some common principles, as exemplified in international human rights instruments and in UNESCO's *Universal Declaration on the Human Genome and Human Rights* (1997), human beings are infinitely complex in their own personal choices. It is in the processes and procedures of the interpretation and implementation of these principles that room must be given to express that complexity.

It is through the personal quest to discover and express this

The tendency to reduce complex issues to black-and-white positions with simplistic media clips borders on the unethical. Human beings are infinitely complex in their own personal choices.

complexity—through information, counselling, and education—that social responsibility takes shape. Frameworks of principles and codes of ethics should lead to the questioning and understanding of how personal responses in the area of genetic choice can, in a cumulative way, affect the family, society, and humanity itself.

The construction of these frameworks is no easy task, especially at the international level. As a member of UNESCO's International Bioethics Committee, I am constantly reminded of how differing world views and cultures—to say nothing of historical and political animosities that often underlie the positions taken on ethical issues—can shape personal convictions. While these world views may slow down the formulation of responses that can be agreed upon, the eventual proposals are all the richer. International harmonization and codification are not for the impatient. They will be all the more difficult to achieve as the genetic revolution continues.

Indeed, an increasing knowledge of the role of genetic factors in common diseases such as cancer, hypertension, and diabetes has a certain effect. It moves us away from a monogenic model of genetic conditions, and toward a multifactorial one. It moves us toward a model where socio-economic factors, as well as environmental and cultural ones, play an important role. Our socio-ethical responses, like those of the political systems that govern how we live and choose, must be equally epigenetic and complex.

Our fear of letting persons choose creates moral lassitude. Ultimately, it also hinders the right of free expression, a right that includes the freedom of research and of scientific endeavour. Interference with that right is justified when the purpose of research infringes on fundamental universal values, such as human dignity and uniqueness—as in the case of human cloning. There will be other potentially dangerous developments, however, that will require extensive and intelligent discussion to reach informed consensus and to formulate sensible laws. Reactive, ad hoc ethics and laws are not in our larger interests, either as individuals or as a society.

Projecting forward, while mindful of the lessons that history teaches us, can trigger the accusation of being speculative. For example, in 1976, when I was studying the ethical, legal, and social issues emerging in new reproductive technologies, I was told that serious scholars did not engage in science fiction. Is it science fiction to ask questions that are grounded in ongoing research, but that try to prepare the parameters of the debate in a thoughtful way that avoids the polemic mentioned earlier? Should we not be encouraging young people to be thoughtful, informed, and engaged? Is that not true "scholarly" activity?

Reactive, ad hoc ethics and laws are not in our larger interests, either as individuals or as a society.

Even more difficult than convincing one's peers that this is all serious business is initiating the discussion at the local, national, and international levels. As we have learned, international discourse is slow and laborious. Ultimately, however, it is enriching, since it provides original and diverse insights for national debates. We can learn from the approaches of other countries, their models for public participation, their mistakes, and their successes in the elaboration of bio-policy.

Here in Canada, the lack of legislation or definitive court decisions does not mean that there are no applicable principles guiding us. Whether the courts address quality of life choices, whether they rule on abortion or the right of a child born disabled to sue its mother, or whether royal commissions and national committees on biotechnology submit reports, the absence of specific law is not the equivalent of no law. A warning to those who would take an interest in these issues: Unlike participation on international committees, working on these Canadian positions attracts media scrutiny that invades one's personal and family life. During my work with the Royal Commission on New Reproductive Technologies, harassment was a fact of life. At this level, what is needed is not only patience, but endurance and the hope that such public service and its eventual product will contribute to advancing tolerant debate and wise decision-making.

If we are to be respectful of genetic and cultural diversity and complexity, and of freedom of expression that allows for personal and responsible decision-making, then we must continue to nourish frameworks of "living" principles. I would argue that classical principles founded on respect for human dignities, such as integrity, privacy, and liberty, need to be reinterpreted in the context of new technology. The juxtaposition of genetics and informatics, of transgenics and virology, of pharmacology and genomics, has created new challenges. We now have the perfect opportunity to resuscitate these classical principles and symbols to give them modern life and personal significance.

In 1976, when I was studying the ethical, legal, and social issues emerging in new reproductive technologies, I was told that serious scholars did not engage in science fiction.

Biologically speaking, we may be just another form of living matter. However, we alone among the Earth's species have the power—and the responsibility—to make sense of our present, and to shape our future. In fact, the "unknown" lies not only in scientific and intellectual discovery, but in something else. It lies in the challenge to discover and create meaningful ethics and policy that will enrich people's lives.

Further Reading

"Commercialization of Genetic Research and Public Policy," Bartha Maria Knoppers, M. Hirtle, and K. Glass, *Science*, Vol. 286, (1999), pp. 2277–78.

"Status, Sale and Patenting of Human Genetic Material: An International Survey," Bartha Maria Knoppers, *Nature Genetics*, Vol. 22 (1999), pp. 23–26.

Report on Genetic Testing, Bartha Maria Knoppers, project director (Ontario Law Reform Commission, 1997).

Human Dignity and Genetic Heritage, Bartha Maria Knoppers (Law Reform Commission of Canada, 1991).

Dr. Robert Gillham received his Ph.D. in soil physics from the University of Illinois in 1972. He has been a professor of hydrogeology in the earth sciences department at the University of Waterloo since 1974, with research interests in the area of contaminant transport and remediation in groundwater.

From 1987 to 1992, Dr. Gillham was the founding director of the Waterloo Centre for Groundwater Research (a Province of Ontario Centre of Excellence), and chair of the Department of Earth Sciences from 1993 to 1997. He currently holds the NSERC/Motorola/EnviroMetal Industrial Research Chair in Groundwater Remediation. Recently, he served as special advisor to the Walkerton Commission and is program leader of the Canadian Water Network.

Among his honours and distinctions, Dr. Gillham has received the Thomas Roy award of the Canadian Geotechnical Society and the Miroslaw Romanowski Medal of the Royal Society of Canada. In 1997, he was inducted into the Royal Society of Canada, and he was granted a D.Sc. by the University of Guelph in 1999. In 2002, Dr. Gillham was made a Member of the Order of Canada. He and his wife, Virginia, live in Guelph, Ontario, and are active in promoting and supporting a number of cultural initiatives.

A New Method for Cleaning Groundwater

Robert Gillham

As one trained in the disciplined and systematic procedures of science, I have found the consequences of a chance occurrence to be both humbling and profound.

In Canada, close to thirty percent of the population relies on groundwater for its domestic water supply. In the United States, that figure is closer to sixty percent. In many countries—Denmark, for example—almost the entire supply of fresh water comes from groundwater. The potential for bacterial contamination of groundwater has been recognized for many decades. Because of the filtering action of soil materials, however, and because of the favourable chemical and biological reactions that can take place within soil materials, groundwater has long been regarded as a safe and secure source of drinking water.

Now we know this is not the case. Our innocence, and that of most of the industrialized world, was severely shaken in the mid-1970s by the tragic events at Love Canal in Niagara Falls, New York. Love Canal, and similar occurrences in the United States, provided irrefutable evidence of the devastation to groundwater that can be caused by the misuse and improper disposal of industrial chemicals.

Public awareness of this issue has been further heightened by the occurrence of groundwater contamination in Woburn, Massachusetts, chronicled in the movie and best-selling book *A Civil Action*. Indeed, a 1994 report of the United States National Research Council, based on extensive studies by various agencies, indicated that there are between 300,000 and 400,000 hazardous-waste sites in the U.S. It further stated that $750 billion would have to be spent in cleaning up groundwater at these sites between 1994 and 2024. These statistics leave little doubt about the vulnerability of groundwater to human activity, and the associated economic liability.

Although comparable statistics are not available for Canada, occurrences of groundwater contamination by industrial chemicals, such as those reported in Ville Mercier in Quebec, and Elmira and Smithville in Ontario, make it clear that the problem does not recognize international boundaries. Indeed, though the contaminants were biological rather than chemical, in the aftermath of the Walkerton tragedy, there should be no doubt at all in the minds of Canadians about the vulnerability of groundwater, and the need for effective policies to protect this vital resource.

Responding to the alarming number of contamination incidents in the early eighties, the United States Congress introduced legislation requiring the cleanup of contaminated groundwater. With the stroke of a pen, a new and substantial industry was spawned: groundwater remediation. Unfortunately, as became recognized over the next decade, and as was documented in the 1994 report of the United States National Research Council, neither the experience base nor the technical knowledge of the problem was sufficient for the industry to meet the challenge.

In the past, the technique most commonly used to remediate contaminated aquifers was to pump the contaminated water to the surface, and then treat it with a variety of methods that would remove the particular contaminants. The process would become

In the aftermath of the Walkerton tragedy, there should be no doubt at all in the minds of Canadians about the vulnerability of groundwater, and the need for effective policies to protect this vital resource.

known as "pump-and-treat." The most commonly encountered contaminants at industrial waste disposal sites are chlorinated solvents. These chemicals are used in almost every industrial sector as cleaning agents, degreasers, and feed stocks for producing other chemicals. Vast quantities have been used in machining and metal plating operations, in the semiconductor and aerospace industries, as well as in dry-cleaning facilities. Many of these chemicals are highly toxic, even at very low concentrations; in groundwater they are highly mobile and can persist in the subsurface for many decades. Furthermore, pump-and-treat has proved to be a particularly ineffective strategy for dealing with these chemicals.

Despite the vast amounts of money directed toward research for the development of more effective remediation methods, little progress had been made by the early nineties. The growing magnitude and severity of the problem, together with the recognition of our technical limitations, contributed to an increasing sense of despair about our ability to solve what had become one of the world's most serious environmental problems.

Then, in 1984, one of my graduate students stumbled on something. The student, who was conducting laboratory experiments that had nothing to do with water contamination, observed the disappearance of several dissolved organic chemicals after they had come into contact with certain metals. Though it was a stimulus for curiosity, this observation was largely ignored at first. In 1989, however, while reviewing the data, I realized what my graduate student had observed—and that enormous efforts were being expended all around the world to achieve what the student had unintentionally achieved five years earlier. The tests were shown to be reproducible, and an extensive series of laboratory tests was undertaken to evaluate the nature of the chemical processes, and the range in applicability of the technology.

What had we found? We found that a wide range of chlorinated organic chemicals commonly found in groundwater could be degraded, by chemical reactions, into harmless products. Because of its reactivity, price, and availability, granular iron—available in large quantities as a byproduct of metal machining operations—was proposed as the most cost-effective reactant. The rates of degradation were generally found to be many thousands of times greater than those observed in the natural subsurface environment. The process was shown to be an oxidation-reduction reaction, with the metallic iron undergoing oxidation as it released electrons, and the organic chemical undergoing reduction as it took on electrons. Through this exchange of electrons, the chemical form of both the iron and the organic chemical were radically changed. The metallic iron was altered to a soluble form (Fe^{2+}), and the chlorine of the organic compound replaced by hydrogen, rendering it harmless. Indeed, the process was not unlike the rusting of iron, with the role of oxygen being replaced by the chlorinated organic chemical.

In the lab we had found the ultimate solution, but how could we apply it in a practical sense? Because the reactions are spontaneous at ambient temperatures and pressures, and because granular iron is the only reactant, it was proposed that the technology be implemented by constructing trenches filled with granular iron across the path of a contaminant plume. As the contaminated water flows through the iron, the contaminants are degraded to harmless compounds and the uncontaminated water is allowed to continue along its natural course. This approach avoids many of the technical problems associated with the pump-and-treat method, and because it is entirely passive, it avoids the large operating and maintenance costs of pump-and-treat systems.

Patents were filed in 1990, and a spin-off company of the University of Waterloo, EnviroMetal Technologies Inc., was formed in 1992 for the purpose of developing and marketing the technology. The technology has been applied or tested at sixty contaminated sites in Canada, the United States, Britain, Europe, and Australia, and is generating economic activity to the tune of

several tens of millions of dollars per year. It is also safe to say that the economy of Canada is benefiting by more than one hundred times the research dollars that were contributed by various government programs. Furthermore, the technology has captured the interest of the international research community, with the literature on the topic having mushroomed from no publications only a decade ago to well in excess of four hundred.

Though the technology has received considerable recognition, it is clearly not high-tech. Depending on one's perspective, it can be viewed as either simplistic or elegant. Sometimes, however, the best solutions are also the simplest—and only require discovery.

In accepting discovery as the inventor, we must also accept the fact that the discovery was not a consequence of deep scientific insight or long hours in the laboratory, but of chance. In fact, little followed the conventional scientific method. While one would never suggest that this discovery falls into the same elite category, there are unmistakable parallels with Newton under the apple tree, Archimedes in the bath tub, Fleming by the open window, and the numerous other serendipitous events that now shape our understanding of the universe. Indeed, one might argue that, while our scientific institutions are well equipped to encourage innovation and the advancement of conventional thought, they are not well structured to encourage discovery.

In accepting discovery as the inventor, we must also accept the fact that the discovery was not a consequence of deep scientific insight or long hours in the laboratory, but of chance.

It is perhaps appropriate to consider the distinction between scientific advancement and discovery. Science generally advances in small steps, with each new investigation building on the results of previous studies, and involving the testing of hypotheses developed from existing knowledge. In this natural course of science, the advances that come to public attention as discoveries are frequently the culmination of a long series of small, logical steps. True discovery, on the other hand, provides new knowledge that can then be explored, enhanced, and developed through conventional science. Fleming's discovery of penicillin, and the subsequent emergence of the science and commerce of antibiotics, is a notable example.

In the situation with which I am most familiar, our initial submission to a refereed journal was rejected. There was not sufficient knowledge or background to write a research proposal that would lead to the discovery, and, had such a proposal been written, there was little likelihood of its being funded. All of this leads to the conclusion that conventional science discourages significant departures from conventional thought, even though chance and serendipity are significant elements in scientific advancement.

Having rationalized and accepted my role as an instrument of chance, I have remained plagued by the five-year lapse between the time the critical observations were made and the significance of the observations were realized. I found little solace in my suspicion that the day Archimedes discovered the laws of buoyancy was probably not the first time he had taken a bath. I have, however, found comfort in these words of Louis Pasteur: "In the field of observation, chance favours only the prepared mind." When I reviewed my graduate student's lab findings in 1989, the needs and problems associated with groundwater cleanup were much more apparent than they were in 1984. As a result, upon review, the implications and potential importance of the data became immediately obvious. Perhaps my mind had been prepared.

Accepting that chance plays a significant role in discovery, and in the advancement of science, gives reason to explore methods for stimulating chance. Having contemplated this question periodically over the past decade, I confess to having few answers. The debate concerning fundamental versus applied research continues, with the assertion that fundamental research is necessary in order to arrive at new understandings and discoveries. Ironically, in this particular case, the experiments that led to the discovery were of a very applied nature, but the discovery has since led to very fundamental studies of charge transfer at corroded metal surfaces, and of chemical pathways and kinetics.

If the answer is not in the fundamental-versus-applied debate, perhaps we should examine our educational and training programs.

It is important that curiosity be stimulated, that students be exposed to a wide range of current scientific thought, and that they graduate with a healthy scepticism toward convention. Perhaps there should be less emphasis on the conveyance of facts, and more on "preparing minds to be favoured by chance."

Further Reading

A Civil Action, Jonathan Harr (New York: Random House, 1995).

Alternatives for Ground Water Cleanup, report of the National Academy of Science Committee on Ground Water Clean-up Alternatives. (Washington, DC: National Academy Press, 1994).

For more information on groundwater remediation, visit EnviroMetal Technologies Inc.'s Web site at: http://www.eti.ca.

Paul Bernard is a professor of sociology at the University of Montreal. His areas of research and teaching include labour markets and social inequality, epistemology, research design, and methods. Dr. Bernard's recent work has focused on economic and labour market segmentation, contingent work, job quality, the living arrangements of young people, welfare and production regimes, and social indicators.

Dr. Bernard has long been passionately involved in the planning and development of social science research and its contribution to shaping social policy. He has been a member of the Social Sciences and Humanities Research Council (SSHRC). He has also participated in many committees at Statistics Canada and sits on the National Statistics Council of Canada. He helped bring about the Data Liberation Initiative and chaired a working group of Statistics Canada and SSHRC that shaped the Canadian Initiative on Social Statistics. As well, he played a key role in convincing the UNESCO Institute for Statistics to move its headquarters to Montreal.

Dr. Bernard now chairs the Research Data Centre's National Coordination Committee. He also serves as chair of the Sub-Committee on Research of the Canadian Sociology and Anthropology Association; is a board member of the Social Research and Development Corporation (which runs large-scale experimental research projects); and is a member of the External Advisory Committee to the Treasury Board Secretary in Ottawa for the annual report, *Canada's Performance*.

Since 1980, Dr. Bernard has authored or co-authored more than sixty books, articles, essays, and policy papers on a wide variety of sociology-related topics. He is a member of the editorial boards of three academic journals: the *Canadian Journal of Sociology*, *Canadian Public Policy*, and *ISUMA-Canadian Journal of Policy Research*.

In 2001, Dr. Bernard received the Outstanding Contribution Award from the Canadian Sociology and Anthropology Association, in recognition of his remarkable contribution to the advancement of sociology in Canada. Dr. Bernard has an M.A. in sociology from the University of Montreal and a Ph.D. in sociology from Harvard University.

The Challenges of Constructing Social Indicators

Paul Bernard

Seldom a day passes without the media bombarding us with the key vital signs of our economy. The unemployment rate. The inflation rate. The Bank of Canada discount rate. The growth rate of the gross domestic product (GDP)—and its per capita level. The state of the balance of payments. Add to that the almighty set of governing rates—TSE, Dow Jones, Nasdaq—that relate to stockmarket performance, and we've painted a picture of a society hanging by a thread of figures and statistics. For all practical purposes, it is as if our very lives depended on such indicators.

We know better than that, of course. For all the enjoyment that money provides in a society where so many things can (and have to) be purchased, and for all the misery that a lack of money, and the means to earn it, can bring, we realize that there is more to life—in particular to life in society—than can be summed up in economic indicators. Along with this realization comes the challenge to construct social indicators that somehow

capture, in a limited set of telling numbers, the key vital signs of our social life.

As part of the challenge, we must start by asking some important questions. For instance, how well are we doing as a society in allowing people, especially children, to develop as human beings? To what extent can people entertain projects and enjoy accomplishments in their work or family life, or through community and public involvement? Are people overstressed by the demands of daily life, as they earn a living and raise children? How much support do they get in these undertakings? Do they get support from family and friends, or from their communities and the state? Are we doing better or worse in these respects than we used to? How do we compare with other advanced societies? How well do vulnerable social groups—the poor, the homeless, women, children, older people, youth, minorities, the disabled—fare within our society?

How can we get a good grasp of the *quality* of social life if we only use numbers—indicators that are best suited to registering *quantity*?

That's not the only challenge, though. There is a second one. How can we get a good grasp of the *quality* of social life if we only use numbers—indicators that are best suited to registering *quantity*? It is a good question. Economists take a straightforward approach because everything of concern to them can be somehow boiled down to amounts of goods and services, and of their universal equivalent, money.

Sociologists see it differently. In trying to develop quantitative social indicators, they hit upon two main difficulties. The first is the role of subjectivity, that is, of the social experiences of different individuals and groups. These are difficult to compare because they are assessed against different sets of expectations. An apparently enviable situation for one person may be deemed unsatisfactory by another who has developed higher preferences—often as a result of enjoying that very situation. On the other hand, deprivation and discrimination may be considered normal by those who have become used to them.

Labour force surveys provide abundant information about the individual characteristics of workers, but very little about the characteristics of the organizations in which they are employed.

Another difficulty comes from the fact that the fabric of social life is composed of social relations, the workings of which are not easily captured with numbers. For instance, let us consider work. Labour force surveys provide abundant information about the individual characteristics of workers, but very little about the characteristics of the organizations in which they are employed. Data collection about these organizations usually resembles a financial balance sheet, with quantities and costs, but with very little information about how work is organized, and with very little about the human aspect. The relationship between workers and workplaces, which involves both conflict and cooperation, and which is so crucial to workers' satisfaction and to firms' performance (as shown by Canadian sociologist Graham S. Lowe in *The Quality of Work: A People-Centred Agenda*), is usually not adequately represented in survey data (although Statistics Canada's recent "Workplace and Employee Survey" will help fill the gap).

The same situation prevails in most areas of social statistics. We have data on the health of people and on how they use health care services. We also have data on health care providers. But we know little of how the two are matched, or of the accessibility, quality, and appropriateness of therapeutic interventions performed in various contexts. The same goes for the education system, the justice system, and so on. In fact, the family is about the only institution where the situation is different. A number of surveys on broad social issues gather data on all members of households so that we can examine the situation of individuals in the context of their daily lives.

This raises more questions. How well are the social sciences facing these challenges? Can they provide decision-makers, and indeed the public, with the social statistics needed for democratic decision-making? And can they provide individuals with the information they need to orient their own destinies, and those of their families?

The real challenges facing advanced societies like Canada concern other factors, such as persistent child poverty in the midst of plenty, employment equity, the integration of young people into adult life, or the inclusion of minority groups and visible minorities.

Consider the much-publicized Human Development Index (HDI) that the United Nations Development Program has proposed yearly over the past decade. It attempts to draw the public's attention to the ends of progress, rather than simply toward one of its *means*—the GDP per capita. The HDI is interesting because it rests on a definite conceptualization of what social life is all about: that societies should allow people to develop fully as human beings. However, this can only happen when people enjoy the energy provided by good health, when they have a roof over their head and some idea of where their next meal is coming from, and when they are empowered—individually and collectively—by access to the treasure of knowledge and culture accumulated by humanity. This approach is based on the concept of capabilities, proposed by the Nobel Prize–winning economist and philosopher Amartya Sen.

As a result, the HDI affirms that some needs are more important than others. That is why income is just one component of the HDI. Health and knowledge, in particular, are considered not so much as *goods* one buys for oneself, but rather as *resources* that members of the community collectively produce and share. This type of index does make a difference. International comparisons reveal that the correlation between affluence and human well-being is far from perfect. In fact, quite a few countries have a high GDP per capita, but score relatively low on the HDI, and vice versa. What does this tell us? The decisive element is not wealth itself, but what nations *make* of their wealth.

The HDI illustrates many of the difficulties involved in constructing synthetic social indicators. In these constructions, three interrelated decisions are being made: decisions about the components of the measure, decisions about the weights attached to these components in the overall index, and decisions about the range

of societies for which the index can legitimately be used as a comparative yardstick.

Some of these difficulties can be seen in Canada. Political discourse here has placed great emphasis on the fact that Canada is repeatedly ranked close to the top of the HDI. As many observers have pointed out, however, it is hardly a feat for our country to be strong on a measure that is more appropriate for determining the relative situation of developing and underdeveloped countries. The real challenges facing advanced societies like Canada concern other factors, such as persistent child poverty in the midst of plenty, employment equity, the integration of young people into adult life, or the inclusion of minority groups and visible minorities.

The other difficulty—the adoption of a system of weights for the various components—has momentous consequences for the ranking of countries. It obviously reflects value preferences rather than scientific choices (so does the selection of the components of indices). Other than admitting this arbitrariness and moving right along, which is done in a surprisingly high proportion of indices, we have two basic options. One is to renounce the attraction of the single-index number and focus, in succession, on a limited number of key dimensions. The other is to use various empirically derived weight systems based on such measuring rods as monetary value, time, or global subjective assessments by the people involved in the situations under study. Empirical correlations among variables can also provide a basis for aggregating them into indices.

Quite a few countries have a high GDP per capita, but score relatively low on the HDI, and vice versa. What does this tell us? The decisive element is not wealth itself, but what nations make of their wealth.

While we're focusing on key dimensions, we may as well keep exploring the three variables featured in the HDI: health, education, and poverty. They can be considered as global-outcome variables because they are so broad as to register just about everything that has happened to people over their lifetime.

Health, for instance, obviously depends on more than just genetics and health care. It reflects one's birth and early childhood circumstances. It reflects education, work experiences, economic situation and trajectory, marital experiences, social support, loves, friendships, and so on. Conversely, health significantly affects all of these other aspects of one's life.

Of course, there are a number of ways to measure health and the incidence of various illnesses. Healthy life expectancy, however, provides a very good overall indication of how life enjoyment is distributed across social classes and around the planet. Like life expectancy itself, healthy life expectancy is closely related to the socio-economic situation of people within societies. What does this mean? It is simple. The richer people are, the longer they live. The richer they are, the more likely they are to enjoy a healthy life for a longer time. On a world scale, the differences are simply stunning. First on the list is Japan, with 74.5 years of healthy life expectancy, and last is Sierra Leone, with about 26 years. Canada places twelfth with 72 years, and the United States twenty-fourth with 70 years.

The richer people are, the longer they live. The richer they are, the more likely they are to enjoy a healthy life for a longer time. On a world scale, the differences are simply stunning.

Concerning access to education and knowledge, we can cite international comparative research on literacy—in which Canada has been a pioneer. In the sixteen to twenty-five age group, our young people rank in the middle of seven Western countries, while the United States ranks almost last. To a large extent, the relatively high scores achieved by many European countries seem to depend on how they instil strong literacy skills in their least-advantaged youth. The youth with the lowest social origins score particularly well in Sweden and the Netherlands, and particularly badly in the United States.

When we consider poverty in advanced societies, we can focus on examining its incidence among children, which is particularly important because their situation is in no way one of their own making. It is also important because the consequences of poverty will follow them for their whole lives. When the United States official poverty line (made comparable internationally) is applied to OECD countries, child poverty ranges from three to five percent in

Scandinavian countries, to thirty to forty percent in the U.K., Italy, and Spain. The United States comes in at fourteen percent, and Canada at nine-and-a-half percent (according to figures available in the *Innocenti Report Card, 2000*, UNICEF's report on child poverty in the world's wealthiest nations).

Research reveals that the stress associated with living at lower levels of income leads to ill health and poor success in learning. For instance, the Canadian Council on Social Development argues that we should aim at equalizing opportunities for our children so that each child has a fair chance to become a successful adult—with good health, good social skills, good learning skills, and good earning skills. However, the National Longitudinal Survey of Children reveals that (after examining two-parent families only) each increment in household income is systematically related to a number of things, including better functional health, lower hyperactivity scores, better acquisition of vocabulary, better math scores, higher participation in sports, and lower dropout and inactivity rates in late adolescence. In other words, the "hidden injuries of class" hurt the development of many of our children. They also hurt our whole society's well-being and productivity.

What are societies to make of these types of results? The simplest use is to consider them alarm bells. For instance, whenever Canada hits a high or low on international scales, we should ask ourselves what it means. The next logical step is to engage in an analysis of the cause—with global-outcome variables as probes. This is exactly what has been done, for example, in the pioneering Canadian work edited by Evans, Barer, and Marmor, *Why Are Some People Healthy and Others Not?* We might need a similarly fresh approach to the fields of education and poverty, especially an approach that brings in variables describing the institutional and community contexts that impinge so strongly on people's trajectories through childhood, school, work, and life.

There are other avenues to aggregating variables into synthetic social indicators. One is to broaden the scope of money as a universal equivalent by attempting to capture aspects of economic well-being that are not actually priced in markets—that should already be the case if we were to pay attention to what economists

Research reveals that the stress associated with living at lower levels of income leads to ill health and poor success in learning.

call "externalities." The Genuine Progress Indicator builds precisely on this idea by subtracting from the level of consumption the environmental and social costs incurred (for instance, the cost of divorce, traffic jams, accidents, and spending on security). The results are impressive. While the U.S. economy was clearly up for the whole latter half of the twentieth century (in terms of GDP per capita), quality of life has broken ranks and been declining steadily since the 1970s.

In at least two ways, time represents another potential measurement standard for the quality of our lives. First, we can calculate expectancies for stays in happy states, such as being alive, in good health, out of poverty, and living in a family whose members do not experience such problems as unemployment, failure in school, and divorce. There is also a second, almost opposite, use of time as an index of quality of life. Instead of examining happy spells and their coincidences, we can look at the extent of conflicts over time allocation, when contradictory demands and commitments can hardly be fitted into one's schedule. Time stress provides not only a measure of individual distress among, for instance, many dual-career couples, single-mother families, students who earn their living, and freelance workers who alternate between overwork and underwork, but also an indicator of our society's successes and failures at producing institutional arrangements that make for more harmonious lives. These arrangements include maternity leave, reasonable work weeks, secure jobs, affordable childcare and elder-care, and so on. Moreover, time stress is likely to lessen civic participation and social capital, thus contributing to a vicious circle of decreasing quality of life.

As was mentioned earlier, subjective assessments of situations are often not trustworthy. Under certain conditions, however, they can be used in the construction of social indicators and can provide a system of weights for the various components of an overall index. Subjective assessments can also be used to provide guidance in the construction of social indicators that make sense to people. As part of their Quality of Life Indicators Project (QOLIP), the

Canadian Policy Research Networks have run a fascinating experiment that gleaned from citizens' deliberations some very interesting information about their lives. This included how time stress hits them, how they see the delicate balance between access and quality in public services, and their frustrations and successes in trying to get their preoccupations across to those who run schools, hospitals, daycare centres, and so on.

The citizens have spoken. Now it's up to the experts to take it from there. They can come up with prototypes of social indicators that will address these preoccupations, test them against expectations, and then venture to provide the people with what they want to be informed about. This is, after all, what constructing social indicators—and indeed policy—is all about.

Further Reading

Why Are Some People Healthy and Others Not? The Determinants of Health of Populations, edited by R.G. Evans, M.L. Barer, and T.R. Marmor (New York: Aldine de Gruyter, 1994).

The Quality of Work: A People-Centred Agenda, Graham S. Lowe (Toronto: Oxford University Press, 2000).

Indicators of Quality of Life in Canada: A Citizens' Prototype, QOLIP project (Canadian Policy Research Networks, April 2001).

"Capability and well-being," Amartya Sen, in *The Quality of Life*, edited by Martha C. Nussbaum and Amartya Sen (Oxford: Clarendon Press, 1993).

Income and Child Well-being: A New Perspective on the Poverty Debate, David P. Ross and Paul Roberts (Ottawa: Canadian Council on Social Development, 1999).

Dr. Wagdi (Fred) Habashi is a professor of mechanical engineering at McGill University. He holds bachelor's and master's degrees in mechanical engineering from McGill, and a Ph.D. in aerospace engineering from Cornell University.

After teaching for two years at the Stevens Institute of Technology in New Jersey, and for twenty-five years at Concordia University, Dr. Habashi established McGill's Computational Fluid Dynamics Laboratory and serves as its director. He is also project leader of the Consortium Laval-UQAM-McGill and Eastern Quebec on supercomputing, which is funded by the Canada Foundation for Innovation and the Quebec government. As well, he is a board member of C3.ca, Canada's coordinating association for supercomputing, is a Fellow of the American Society of Mechanical Engineers, and an Associate Fellow of the American Institute of Aeronautics and Astronautics.

Dr. Habashi maintains a particularly strong interaction with industry, and has been Pratt & Whitney Canada's consultant for the aerodynamics of jet engines since 1977. He was given the additional title of Research Fellow of P&WC in 2000. He has been a consultant for a number of other companies, including Alcan International and Bombardier Aerospace. In 2000, he was awarded the J. Armand Bombardier Chair of Multi-disciplinary CFD at McGill University, in which Silicon Graphics is also a financial participant.

Dr. Habashi is also the founder of three successful spin-off companies: Newmerical Technologies International and Real Numerix Limited, both involved in multi-disciplinary applications of computational fluid dynamics; and Scientific Aircraft Accident Analysis, which is introducing science as a forensic tool in aircraft accident investigation.

His numerous scientific and technology-transfer awards include the E.W.R. Steacie Fellowship from NSERC, the Technology Achievement Award of Pratt & Whitney Canada, the University Research Award of Concordia University, and the Cray Gigaflop Award for the fastest computer code in the world in 1990.

Dr. Habashi is the author of some two hundred journal articles, conference papers, and books, and is editor-in-chief of the *International Journal of Computational Fluid Dynamics*.

Putting Supercomputers on Ice

Wagdi G. Habashi

When supercooled water droplets in clouds hit an aircraft, they create an ice layer with a roughness and form that can lead to substantial distortions in the aerodynamic profiles of the plane's wings, air intakes, and propellers.

This has dangerous implications. Performance degradation can then be caused by a combination of increased drag (as a result of roughness and flow separation) and a reduced stall angle of attack. The increased and shifted weight are additional issues. Asymmetrical roughness distributions can also cause significant stability and control problems, compounding the already reduced aircraft performance. Ice can block engine inlets and internal ducts, and can damage components if it gets sucked in, causing problems such as power fluctuations, thrust loss, rollback, flameout, and loss of transient capability. But stall is the killer. Today's stall protection systems cannot alert the pilot that the margin between a stall warning and an actual stall has been significantly

reduced, and perhaps completely eliminated, in an icing situation.

Ice accretion during flight can be prevented by adding energy in the form of heat (thermal anti-icing), to prevent water droplets from freezing. Accreted ice can be cyclically removed during flight using systems operated by pneumatics (mechanical de-icing), which break the bond at the ice-surface interface, or by thermal de-icing, which melts the ice by internally pushing hot air onto the leading edge of the wings. Unfortunately, the complete prevention of ice formation, or its complete removal, is not, and can never be, economically feasible. This is because of the large amount of thermal or mechanical energy required. Moreover, the controlled amount of anti- or de-icing hot air bled from the engines is often needed during climb, especially for smaller airplanes. In practice, therefore, while some areas of the aircraft are anti-iced, others can only be de-iced, and a large part is left unprotected. Such unprotected areas must be precisely determined and the aircraft must be tested—either behind an icing tanker in an icing tunnel, or through flight in natural icing conditions—before being certified for "flying into known icing."

An ironclad solution against icing has been prevented by two shortcomings: the difficulty of detecting and measuring ice accretion, and the necessarily cyclic nature of de-icing an aircraft during flight. Ice detection systems are installed on only a fraction of the airplanes operating today, and are known to be inefficient. Pilots do not trust ice detection systems and simply monitor places where ice accretes. "If I have ice on the windshield wiper bolt," they reason, "I have ice on the wings." In the case of booted airplanes, the pilot must wait for some ice to accrete before activating the pneumatic de-icing boots. One would think that the precise amount of ice permitted to accrete would be based on scientific principles. It turns out, however, that these decisions are based on no more than a rule of thumb (quarter inch, half an inch) depending on the mechanical ability of the boot to "crack" the ice. Half an inch of ice would have vastly different aerodynamic effects on various aircraft. But can a pilot really sense half an inch of ice on a wing hidden from view, especially at night?

A second shortcoming of de-icing systems is that available

Pilots do not trust ice detection systems and simply monitor places where ice accretes. "If I have ice on the windshield wiper bolt," they reason, "I have ice on the wings."

power dictates that in-flight de-icing operations be cyclic—say wing, tail, empennage, and back—with blackout periods for each component. It makes sense for the wing to be designed to aerodynamically sustain the inter-cycle ice load that accretes during the wing de-icing blackout period, but this is just now being studied.

With major aircraft manufacturers, certification agencies, and national research agencies linked with research developments in the area of in-flight icing, it is only natural for the public to assume that this aspect of flying has been completely mastered. While these entities are certainly trying their best, the fact is that aircraft, system design, and operational procedures still have not totally conquered the in-flight icing problem. Flying in icing conditions continues to result in incidents and accidents, with no type, size, or configuration of aircraft being immune.

I became interested in icing research when a good friend (and now research partner), Captain Gary Wagner of Air Canada, brought the state of icing prevention to my attention, and encouraged me to start thinking of applying my brand of computational fluid dynamics (CFD) and computer simulations to icing situations. We decided to organize a one-day workshop on the matter, to which we invited a "who's who" of icing research. The icing manager at NASA, the U.S. space agency, appeared as our keynote speaker. This address was followed by a three-day course, which we, the workshop organizers, attended as students. The course was given by NASA and ONERA (Office National d'Études et de Recherche Aérospatiale), the custodians of the two most advanced icing simulation codes. We were taken aback that the "big boys" were still using ten- to twenty-year-old computational algorithms and simulation techniques that we had stopped teaching some time ago. The majority of ice accretion codes are two-dimensional and, hence, not geometrically realistic. They use panel methods, and so are limited to incompressible and attached flows, and are based on Lagrangian particle-tracking models and one-dimensional flux-balancing techniques that were dictated by the computer limitations of a decade ago.

We were taken aback that the "big boys" were still using ten- to twenty-year-old computational algorithms and simulation techniques that we had stopped teaching some time ago.

These self-imposed limitations force CFD to play only a token role in icing design and certification. In fact, CFD is limited to predicting ice accretion shapes that are then manufactured from foam and affixed to the aircraft being certified. Performance degradation and airframe control behaviour can then be characterized through flight tests, and in wind or icing tunnels, thus providing the designers with information for the development of control response in such situations.

This limited application of CFD does not take advantage of its ability to predict aerodynamic performance, to a reasonable and often great degree of accuracy, well before any part of an airplane or jet engine is ever built. This self-imposed restriction on the use of scientific tools makes exchanges between aerodynamicists and icing protection engineers, to say the least, painful. Ice protection is often relegated to dealing with an already frozen aerodynamic configuration.

In addition, national certification agencies have made icing simulation software available for free to local industry, and given the impression that using this software may get an aircraft certified more quickly. By doing so, these agencies have created a captive market, and blunted the need for industry to make a major investment in developing its own icing code. Icing simulation software has been frozen in a time warp, under-utilizing available computational power. Canadian aerospace manufacturers have no access to foreign icing codes protected by national interests.

At the Computational Fluid Dynamics Laboratory at McGill University, our intent is to design and develop an icing certification tool that can accurately predict ice accretion on an entire aircraft, under all atmospheric conditions. This will enable designers to predict such details as droplet trajectories, limits of impingement, ice accretion shapes, and melted ice runback, as well as the iced aircraft's performance characteristics. Computer modelling remains the best option for safely testing all possible corners of the envelope. CFD offers extremely accurate modelling, because it respects

geometric fidelity. It also can handle aerodynamic complexity, and presents no physical danger. CFD is the only technology that is multi-disciplinary, reproducible, traceable, upgradable, and continuously decreasing in cost.

We plan to develop the most advanced mathematical tools for analysing aircraft icing, and to provide industry with a "numerical simulator" that can improve and optimize designs, reduce testing, accelerate certification, enable the investigation of situations difficult to reproduce, and provide a more realistic training tool for pilots. We are building a comprehensive CFD tool, named FENSAP-ICE (Finite Element Navier-Stokes Analysis Package for Icing), that is introducing novel applied mathematical concepts, algorithmic solution methods, and new physical models at all levels of icing simulation. It is three-dimensional and CAD-based, and is built in a modular form so that its components can be interfaced with any existing CFD aerodynamics code. Its software design will minimize user intervention by employing grids that move to hug the changing shape of the aircraft while icing and during mesh optimization, thus providing the most cost-effective CFD answer.

Icing simulation software has been frozen in a time warp, under-utilizing available computational power. Canadian aerospace manufacturers have no access to foreign icing codes protected by national interests.

Could FENSAP-ICE be developed without a supercomputer? The answer is a resounding no. Will the necessity for a supercomputer hinder the use of FENSAP-ICE in industry? The answer is another resounding no. The aerospace industry has many supercomputers and, in this way, is second only to the automotive industry (which has been more forthcoming in the acquisition of the needed computer power, mostly for structural and crash analysis).

Yet significant computer resources will be needed. Truly three-dimensional solutions can be carried out on workstations today, but the three-dimensional Navier-Stokes solutions required for heavy CFD calculations almost always require a parallel computer. There are some orders of magnitude to be considered. First, to solve the three-dimensional turbulent flow equations over a complete aircraft may require from five to ten million grid points and, with adaptation, may take from ten to twenty hours on a powerful

64-bit symmetric multiprocessor (64-cpu SMP) computer. Second, it is estimated that a full conjugate heat transfer analysis of a wing under icing conditions would require ten to twelve hours on the same machine, and about three to four hours on a much more powerful, 512-cpu Beowulf cluster. Third, it is estimated that the unsteady calculation of viscous flow over an airplane, under icing conditions, in pitch and yaw, would require 90 to 110 gigabytes of memory, and an execution time of thirty to sixty hours on a 64-cpu SMP machine. Clearly, there is no hope for such methods to be developed without the availability of a supercomputer.

What is the anticipated impact of FENSAP-ICE? It will permit the industry to develop aerodynamic profiles with gracefully degrading aerodynamic performance characteristics that are not as abruptly, or severely, affected by ice. It will permit the industry to design de-icing and anti-icing systems of minimum weight and power with highest efficiency, and to design ice protection systems that are still adequate for meteorological icing conditions even worse than the conditions that must be met for certification. FENSAP-ICE will allow the positioning and design of more effective ice detection systems, and will permit the industry to build aircraft simulators to better train pilots and give them a more realistic rendition of aerodynamic behaviour under icing or iced conditions.

FENSAP-ICE will also make it possible to integrate icing and aerodynamic design. Historically, aerodynamic component design cycles have focused exclusively on performance, dynamics, and structural integrity. Geometry definition, analysis, and testing activities have been tailored to meet the performance objectives. In such a scenario, icing is relegated to the certification test program. Icing-related problems, however, result in substantial changes to the same aerodynamic surfaces that have been so painstakingly refined.

Concurrent design of ice-protected aerodynamic surfaces would allow for a multidisciplinary optimization of all design aspects. Aerodynamic performance, within the constraints of adequate ice protection, would be a focus of each step in the design cycle—including geometry definition, aerodynamic icing analysis, and testing. In such a scenario, icing certification would not be coupled

with unplanned protection system development. Instead, product definition would be complete prior to compliance testing.

FENSAP-ICE will also permit manufacturers to consider certification requirements as a baseline (or minimum) requirement, and to go beyond them. For example, recent air accidents have shown that it may be necessary to consider situations exceeding those currently specified in certification rules, such as supercooled large droplet (SLD) icing. This presents difficulties, since current meteorological techniques do not readily predict these droplets, and available measurement devices cannot adequately sense them. CFD, however, can simulate them. By putting supercomputers on ice, Canada is guaranteeing itself a technological lead for the aircraft it manufactures. It is also ensuring greater safety for the flying public.

Further Reading

"Flying into trouble," Mary Schiavo, *Time* (Canadian edition), Vol. 149, No. 13 (1997), pp. 34–44.

"Putting Computers on Ice," Wagdi G. Habashi, ICAO *Journal*, Vol. 50, No. 7 (1995), pp. 14–17.

"Pilots can minimize the likelihood of aircraft roll upset in severe icing," J.P. Dow, Sr., *Flight Safety Digest* (January 1996), pp. 1–9.

"Advancing icing detection," C.C. Ryerson and G.G. Koenig, *Aerospace Engineering* (November 2000), pp. 22–24.

"Takeoff and landing in icing conditions," G.A. Wagner, *Pilot* (December 1989); reprinted by *Airline Pilot*, magazine of US ALPA, Washington, DC, (November 1990); reprinted by *Flight Deck*, British Airways safety publication, Issue 14 (Winter 1994–95).

"Effects of adverse weather on aerodynamics" (Neuilly-Sur-Seine: Advisory Group for Aerospace Research and Development, December 1991), p. 496.

Linda Hutcheon is a professor of English and comparative literature at the University of Toronto. Her many books on literary subjects have established her as a major literary theorist in North America, and have helped to promote a greater understanding of modern fiction, parody, postmodern literature, irony, feminist theory, and ethnic minority writing in Canada.

Dr. Hutcheon attended the University of Toronto, where she earned an honours B.A. in modern languages in 1969. In 1971, she obtained an M.A. in Romance studies from Cornell University, and then returned to the University of Toronto, where she earned a Ph.D. in comparative literature in 1975. She taught English for twelve years at McMaster University, at which time she began to publish her books on literary theory. She has also been a visiting professor at a number of universities in Italy, Puerto Rico, and the U.S., and has lectured across North America and Europe. She was awarded a Guggenheim Fellowship, as well as honorary doctorates from Concordia University, the University of Western Ontario, and the University of Antwerp.

Dr. Hutcheon is the holder of Woodrow Wilson, Killam (Junior and Senior), Connaught, and Social Sciences and Humanities Research Council (SSHRC) research grants. She has just completed her term as the 110th president of the Modern Languages Association of America—one of the largest humanities organizations in the world. She is also a member of the SSHRC Working Group on the Future of the Humanities, and is currently supervising seventeen doctoral dissertations and two postdoctoral projects. In addition to the comparative study described here, she is involved in another collaborative interdisciplinary project (funded by SSHRC)—with Michael Hutcheon, and both graduate and undergraduate research assistants—on the topic of death and dying, once again using opera as the vehicle for studying cultural and social history.

Dr. Hutcheon has published eight single-authored and two co-authored books of cultural commentary and criticism: *Narcissistic Narrative* (1980; rpt. 1984); *Formalism and the Freudian Aesthetic* (1984); *A Theory of Parody* (1985; rpt. 2000); *The Poetics of Postmodernism: History, Theory, Fiction* (1988); *The Canadian Postmodern* (1988); *The Politics of Postmodernism* (1989; rpt. 2002); *Splitting Images: Contemporary Canadian Ironies* (1991); *Irony's Edge* (1995); and, with Michael Hutcheon, M.D., *Opera: Desire, Disease, Death* (1996) and *Bodily Charm: Living Opera* (2000). Her books have been translated into Japanese, Chinese, Portuguese, Korean, Serbian, and Italian.

Culture on the Move: Rethinking Literature and Its History

Linda Hutcheon

As twenty-first-century Canadians living in a global world, we know that culture travels. It does not stand still. Canada's verbal, visual, and musical arts not only have a clear international impact, but the arts of other cultures inevitably also influence our own. Gone are the days when we would even contemplate defining "Canadian" in anything resembling purist cultural terms—of either a single or double ethnicity.

This realization—that the complex and mobile multiplicity that defines Canadian culture is shared by all the world's cultures today—has allowed us to see our surroundings in a new light. It has also been the starting point for the research on a project entitled "Rethinking Literary History Comparatively" carried out at the University of Toronto from 1996 to 2001 and co-directed by Mario J. Valdés and myself. Funded by the university and the Social Sciences and Humanities Research Council, this five-year, five-volume project involved hundreds of scholars from around the

world working together with one common goal: to retell the story of the literature of two of the more culturally complex parts of the world—Latin America and East Central Europe.

Since the nineteenth century, the story of literature's past has traditionally been told in one of two ways. It was told either in terms of the nation, as in the *Literary History of Canada*, or in terms of the national language, as in *A New History of French Literature*. But as the twenty-first century approached, and the realities of the globalization of culture became increasingly apparent, the nineties seemed an appropriate time to reconsider long-held beliefs about the way to write literary history. As a result, in July 1993, Mario Valdés brought together a team of twenty-two scholars from nineteen institutions in ten countries. Their mission was to begin the rethinking of how we might better tell the story of literature's past—as a way to understand its present.

The group of scholars first met at the Rockefeller Foundation's Bellagio Study and Conference Center in Bellagio, Italy. From those initial, lively discussions came the theoretical and historical frameworks for the various parts of the project, as were outlined in a series of published works (see Further Reading).

Instead of simply accepting the standard national model for literary history, in Bellagio we found ourselves starting from scratch. We believed the problem lay in the fact that the traditional model was usually premised on the convenient fiction of a single ethnicity and a single language. It presented the past of a nation's literature from recognizable, "primordial" origins, through to a present that has developed organically from those roots. It did so as a way to project its future through to the inevitable *telos*, or end point, that authorizes its cultural value. The basic concepts behind this model, as well as its general shape, developed out of a heady blend of nineteenth-century German Romantic idealism, European cultural nationalism, and the rise of philology (the linguistic discipline that did so much to establish the Western sense of the specificity of languages, and of the peoples and nations speaking them). With its emphasis on the importance of origins and the assumption of continuous organic development, this teleological model was

intended to establish an implicit parallel between the inevitable progress of the nation and its literature. Historically, the modern nation-state and the discipline of literary history, as we know it today, were indeed born together in the nineteenth century. They have, therefore, been mutually implicated from the start.

However, given the major shifts in world culture—thanks to electronic technology and global capitalism—it was clearly time for a change. The first step was to jettison the limitation of the single nation. In the Latin American project, a new comparative model was developed and elaborated with the aid of cultural geographers. The goal was to map the movement of culture across national and linguistic boundaries, tracing the vast continent's verbal cultures on their travels from being aboriginal societies, through to the European conquest, and up to the present day. The term "verbal culture" itself marks the second step, which was a widening of the definition of what is considered "literary" to include such things as indigenous narratives, travel accounts, sermons, *testimonio*, film, opera, journalism, and many other verbal forms not traditionally considered to be part of "literature." In short, in both parts of the project, we extended the boundaries of the literary from the earlier, restricted notion of imaginative writing to include many other categories—documentary as well as fictional, oral as well as written, popular as well as elite.

Cultures travelling. Cultures coming together. Only by working in these more comparative and inclusive ways did we feel we could avoid the kind of sectarian thinking that, at its worst today, leads to the violent nationalist eruptions we are still witnessing the world over.

In the East Central European volume, it was a concept of cultural "nodes"—where cultures overlap and come together—that was used to structure the narrative of this region's complex past. Sometimes these nodes were geographical entities, such as the River Danube, that allowed cultural features not only to travel, but to merge and mutate. Sometimes they were people, like the Jewish Franz Kafka, writing in German in Prague. Sometimes they were places—like Gdansk/Danzig—whose names may even have changed over the years of political and cultural border shifting.

Literary histories confer cultural authority and create a sense of continuity between past and present—usually with an eye to promoting some kind of ideological consensus. In short, they are not innocent documents.

Now we arrive at the third step in the rethinking of literary history: a consideration of what has come to be called the "literary institution." By this, we mean the broader economic, political, and social field in which the experience of literature takes place. The field in which readers read and writers write. Clearly, these new literary histories, built on these different models, are not going to read like traditional teleological ones. Their very forms question, both implicitly and explicitly, the basic assumptions of historical narrative, including those of time frame and of the authority of the historian's perspective.

Cultures travelling. Cultures coming together. Only by working in these more comparative and inclusive ways did we feel we could avoid the kind of sectarian thinking that, at its worst today, leads to the violent nationalist eruptions we are still witnessing the world over. Yet, the nation is no longer the dominant defining power in the world of transnational capitalism. With people everywhere on the move, by choice or by force, the diasporic dimensions of globalization have undermined the coherence of the concept—not to mention the power—of the nation-state. We felt that all these changes required parallel changes in how we think about, and how we shape, literary historical narratives. Literary histories confer cultural authority and create a sense of continuity between past and present—usually with an eye to promoting some kind of ideological consensus. In short, they are not innocent documents.

Any literary history, therefore, runs the risk of excluding or marginalizing works that do not fit its model. However, the more flexible, self-conscious, and integrative the model is, the greater the chance of its being inclusive. For instance, as Canadians know, a single-language, single-ethnicity national model has difficulty dealing with the obvious fact that people can, and often do, participate in several language communities at once. Texts, as well as ideas and images, pass from one language to another through

the medium of translation. Films are released in many languages at the same time. Novels are simultaneously published around the world in translation. Plays are performed on several continents in different languages. What we believed we needed for our new literary histories were models that took this kind of diversity and complexity into account.

To accomplish this, we involved not only literary historians, but also scholars from a variety of different disciplines including geography, history, sociology, anthropology, economics, linguistics, musicology, film, and visual art studies. Needless to say, the collaborative nature of the project was inevitable, given the extended range of materials to be considered. That hundreds of contributors were willing and able to drop everything, and to immediately come on board a time-limited project like this, suggests something significant; we were not alone in believing that the time was ripe to rethink literary history. That this kind of collaborative work is really only possible in our current age of electronic communication likely goes without saying. I should note, however, the crucial fact that our contributors lived in many nations and worked under radically different research conditions.

We hope that the cumulative effect of this progressive enlarging of the field of literary history will be one of the most important scholarly contributions of this project. The specific contributions included its expansion of the geographical/political scope beyond the nation-state; its elasticizing of the sense of the literary; its broadening of the focus to include the social and institutional context of reading and writing; its productive challenge to traditional ways of writing history; its embracing of the insights and models of other disciplines. Of course, there were also the collaborative and cooperative working relationships it engendered among scholars from all over the world.

We also hope these scholarly innovations will have an effect on the thinking of readers, both inside and outside the academy. If literary histories can work to create a sense of belonging and recognition for a people—and they certainly do—they can also work to enlarge our sense of what it is that we belong to, and to recognize ourselves as part of it. As citizens of a bilingual and multicultural Canada, we are perhaps especially sensitive to issues of national identity and culture's role in its promotion. We know that how we think about the culture of the past cannot really be separated from how we act in the present. Perhaps this, above all, is why this had to be a Canadian project.

Further Reading

Oxford University Press has published a series of books that resulted from the 1993 discussions in Bellagio, Italy:

The Oxford Comparative History of Latin American Literary Cultures, edited by Mario Valdés and Djelal Kadir.

The Comparative History of East Central European Literary Cultures: Nineteenth and Twentieth Centuries, edited by Marcel Cornis-Pope, George Grabowicz, Alexander Kiossev, Tomislav Longinovic, John Neubauer, and Svetlana Slapsak.

Rethinking Literary History: A Forum on Theory, with extended essays by Mario J. Valdés, Homi Bhabha, Stephen Greenblatt, Marshall Brown, Walter Mignolo, and Linda Hutcheon.

Also:

Is Literary History Possible? David Perkins (Baltimore: Johns Hopkins University Press, 1992).

The Uses of Literary History, edited by Marshall Brown (Durham, NC: Duke University Press, 1995).

National Culture and the New Global System, Frederick Buell (Baltimore: Johns Hopkins University Press, 1994).

"Rethinking Literary History—Comparatively," Linda Hutcheon and Mario J. Valdés, American Council of Learned Societies Occasional Paper 27 (1995).

"Collaborative Historiography: A Comparative Literary History of Latin America," Linda Hutcheon, Djelal Kadir, and Mario J. Valdés, American Council of Learned Societies Occasional Paper 35 (1996).

Michele Mosca has made major contributions to the theory and practice of quantum information processing. Among his accomplishments, he is credited with clarifying, generalizing, and unifying quantum algorithms; co-inventing the polynomial method for studying the limitations of quantum computers; and co-developing the theories of quantum self-testing and of private quantum channels.

Together with collaborators at Oxford University, Mosca realized several of the first implementations of quantum algorithms, and from 1998 to 1999 was the Oxford project manager for the largest European collaborative grant in experimental quantum cryptography.

He obtained a bachelor's degree in mathematics from the University of Waterloo (through St. Jerome's University) in 1995, and was the recipient of the University of Waterloo Mathematics Faculty Alumni Gold Medal. He also went to Wolfson College at Oxford, on a Commonwealth Scholarship, and received a master's degree in mathematics and the foundations of computer science in 1996.

Mosca continued his education at Oxford University on a U.K. Communications-Electronic Security Group scholarship, obtaining his D.Phil. in quantum computer algorithms in 1999 while holding the Robin Gandy Junior Research Fellowship. Since 1999, he has been assistant professor of mathematics at St. Jerome's University and in the Combinatorics and Optimization department of the Faculty of Mathematics at the University of Waterloo. He is also a member of the Centre for Applied Cryptographic Research, teaches combinatorics and quantum computation, and holds a Premier's Research Excellence Award (2000–05).

Mosca is co-founder and deputy director of the Institute for Quantum Computing at the University of Waterloo, and is the project leader of a Canada-wide NSERC Collaborative Research Opportunities project (with Calgary, McGill, and Montreal) that promotes collaboration with partners in Europe and the U.S. He has also been a government consultant on quantum information technology.

As part of his overall goals, Mosca would like to help Canada maintain and promote its position at the forefront of the field of quantum information processing.

The Promise of Quantum Computing

Michele Mosca

Until now, the information that has become so vital to our modern technological existence has been stored and handled by physical means. No exceptions allowed. Just like everything else on the planet, the techniques we have developed for storing, processing, communicating, and encrypting this information have been obligated to respect the laws of physics, but a subtle shift could be taking place. The most modern and current of these information-processing technologies are starting to push the boundaries of these laws. They may even be creating a few laws of their own.

In reality, the newest of these technologies are quickly approaching a scale so small that the laws of classical physics will no longer be precise enough to correctly describe their behaviour. Increasingly, we are finding that we must turn to the physical theory of quantum mechanics—or the mechanics of the *small*—to explore a qualitatively different and more powerful type of information processing. We generally refer to this process as "quantum computation."

Conventional computing has been based on the theories of classical physics, the "one-two" of its binary code corresponding to the "on/off" of an electrical switch. Quantum computing is based on how nature works at the atomic level.

Why give so much attention to something so small? The remarkable advances in information technology during the last century—mobile phones, global positioning systems, handheld computers, atomic clocks—were built upon a greater understanding of physics and a better control over physical systems. A deeper understanding of quantum physics, combined with new technologies that give us sufficient control over quantum systems, will increase our understanding of the general laws of physics. They will also open the door to a new world of technology and its potential applications. Most of this will be as unimaginable to us as handheld computers and the Internet would have been to Charles Babbage, the nineteenth-century pioneer of computing.

Now some background. Conventional computing has been based on the theories of classical physics, the "one-two" of its binary code corresponding to the "on/off" of an electrical switch. Quantum computing is based on how nature works at the atomic level. A quantum computer may consist, for example, of a collection of hydrogen atoms, in which the code is spelled out by the orientation of the magnetic spin of the proton (the orientation can be changed with radio frequency pulses) or by whether the electrons are circling the atom in either of two patterns. Because the two patterns of each electron can apparently exist together at once, and because each added electron doubles the capacity of the quantum computer, a collection of fifty or a hundred electrons can increase computing power trillions of trillions of times.

How did I first come to the land of quantum computation? Not in a straight line (the shortest distance between two points). I came to it through the side door of cryptography—the art of ensuring the security of information by using mathematical techniques. Anyone who has ever sent personal information out into cyberspace knows that cryptography plays a big role in verifying identity and keeping information such as credit card numbers secret. Much like the rest of us, it relies on "keys" to keep things safe and secure. There are two types of cryptography: private-key and public-key.

In private-key cryptography, two parties (traditionally referred to as "Alice" and "Bob") share a private piece of information called a "key." The key is used with a procedure (protocol) to scramble (encrypt) and unscramble (decrypt) the information. There are countless protocols for encrypting and decrypting information. If Alice and Bob use the famous "one-time" pad protocol (the key can only be used once—hence the name), the encryption is unconditionally secure. Without the key, an adversary has no chance of getting any information about the message—regardless of their computing power. Even if the adversary uses one of the more efficient protocols, which use much smaller, reusable keys, the encryption is believed to be computationally secure. Although, in principle, the code could be cracked, it would take an impractical amount of time.

How did I first come to the land of quantum computation? Not in a straight line (the shortest distance between two points). I came to it through the side door of cryptography—the art of ensuring the security of information by using mathematical techniques.

The main challenge in private-key cryptography is distributing the secret keys. How can two parties exchange keys if they have only met by phone or through the Internet? If Alice makes up a key and sends it through an unsecured channel (such as by e-mail or telephone), then a competent eavesdropper can easily observe the key, make a copy of it, and later use it to decipher any messages that Alice and Bob send to each other. That presents a whole set of security-related problems.

How do we deal with them? With something known as public-key cryptography. The main idea behind this type of cryptography is that each person has two corresponding keys: one encryption

> **I studied the security of a widely used class of cryptographic techniques, and made a discovery. I learned that, unlike classical computers, quantum computers could crack these codes using very few steps.**

key and one decryption key. Each person makes the encryption key (or public key) publicly available and keeps the decryption key secret, for themselves. Only the secret decryption key can decipher messages encrypted with the corresponding encryption key. Even someone who knows the encryption key cannot decrypt any messages. In a nutshell, what does this mean? Just imagine having an open padlock. It can be locked by anyone, but only unlocked with the right key. Many Web browsers today already take care of managing these public and private keys for their users.

Why are we so sure the system is secure? In principle, given Alice's public key, an adversary could eventually compute her private key and decipher any messages intended for her eyes only. Mind you, performing that computation would require trillions of trillions of computational steps. Even if the adversary used the latest technologies, it would still take millions of years. Not the best use of resources by any account. Public-key cryptography, therefore, offers computational security. Does this sound nice and safe? Yes, but a word of warning: The people who make and break codes are always aware that somewhere, somehow, someone might stumble on a shortcut to cracking a code.

That's exactly what happened in the early 1980s when the Data Encryption Group at the University of Waterloo cracked one of the earliest public-key cryptosystems being produced for commercial use. Although their actions might have seemed destructive and "adversarial," the group members were actually the good guys. That's because they showed the industry that the systems they had in place were already unsafe, and ready to be cracked by any competent adversary. The industry took notice and modified their protocols appropriately. As a student, I worked for two members of the Waterloo Data Encryption Group, Ron Mullin and Scott Vanstone, and studied potential weaknesses in cryptosystems in use or being considered for use in industry. Over the past twenty years, the techniques used in industry have been scrutinized by experts all over the world. We believe they are now secure against any realistic mathematical attack.

In 1996, I found myself at Oxford University, continuing my studies in mathematics, this time as they related to information technology. For my M.Sc. dissertation, I studied the security of a widely used class of cryptographic techniques, and made a discovery. I learned that, unlike classical computers, quantum computers could crack these codes using very few steps. Furthermore, I learned that great advances had been made in the development of quantum computers.

When the Nobel laureate Richard Feynman first suggested the idea of a quantum computer in 1982, he pointed out that simulating quantum physics on a classical computer would be very difficult. He believed, however, that a computer built with quantum components could probably perform such simulations efficiently. It was a quantum leap in the right direction. Then, in 1985, Oxford physicist David Deutsch defined the quantum computer, and described a small problem that a quantum computer algorithm could solve faster than any classical computer. Several researchers, including Canadian computer scientist Daniel Simon, then a postdoctoral fellow at the University of Montreal, generalized this tiny quantum algorithm. Shortly after, building on the work of Simon, Peter Shor at AT&T made another remarkable leap and showed how quantum computers could efficiently solve several difficult mathematical problems, including those used in public-key cryptography (namely, factoring large integers and finding discrete logarithms).

The people who make and break codes are always aware that somewhere, somehow, someone might stumble on a shortcut to cracking a code.

Today, researchers around the world and here in Canada—at McGill University and the universities of Calgary, Montreal, and Waterloo—are studying the algorithmic powers and limitations of this new computing mode and looking for ways to solve currently intractable problems.

One of the important properties of quantum information is that there are fundamental limitations to how much information can be extracted from a quantum system. To understand this better, we just have to consider the Heisenberg Uncertainty Principle, which states that when we measure one property, we inevitably perturb the system and lose the ability to measure certain other properties. This means that quantum information is sensitive to eavesdropping. Charles Bennett (IBM) and Gilles Brassard (Montreal) have shown how this sensitivity to eavesdropping could be exploited in order to solve the key-distribution problem identified earlier. If Alice sends Bob some random information encoded in a quantum medium, and an eavesdropper (we'll call her "Eve") tries to look at part of the key, Alice and Bob (after some public discussion about parts of the key) are likely to notice the disturbance Eve caused to the key. If the disturbance is not too great, Alice and Bob can distill a shorter secret key.

In the past one hundred years, great advances have been made. We have moved from observing quantum phenomena to controlling them. We can now trap and observe a single atom, and we can prepare and measure a single particle of light.

To solve this problem, quantum cryptography was born, offering a method of distributing keys that is unconditionally secure. Since then, several research groups around the world have implemented quantum-key exchanges by sending single photons of light through fibre-optic cables over distances of thirty centimetres to nearly sixty kilometres. Some groups are also working toward performing quantum-key exchanges with satellites in orbit by sending photons through the atmosphere.

Researchers worldwide are also working on further applications of quantum mechanics for information security. For example, at the University of Waterloo, Alain Tapp and I, working with collaborators at the University of California, Berkeley, and Amsterdam University, have shown how to encrypt quantum information using classical keys. These keys are easy to maintain, since they can be stored on any traditional medium such as paper, a floppy disk, or a hard drive. We have also proven that our methods make optimal use of the key.

The question remains, however, is quantum computation technologically feasible? Quantum computers do seem to be a realistic model of computation. The obstacles to performing large-scale quantum information processing appear to be only technological, not fundamental.

In the past one hundred years, great advances have been made. We have moved from observing quantum phenomena to controlling them. We can now trap and observe a single atom, and we can prepare and measure a single particle of light. For decades, we have controlled and measured the interaction of quantum particles using nuclear magnetic resonance technology—the same technology used in magnetic resonance imaging (MRI) scans in hospitals. Today, the state-of-the-art is the 7 quantum-bit (or "qubit") computer implemented using NMR technologies by two independent North American research groups (including Canadian Raymond Laflamme, at the University of Waterloo and the Perimeter Institute for Theoretical Physics).

When I first heard about quantum computers, there were strong warnings from scientists, including Rolf Landauer at IBM and William Unruh at UBC, about the sensitivity of quantum systems to environmental disturbances. In 1995 and 1996, however, researchers at AT&T and Oxford University showed how the error-correcting procedures that allow CD players to play scratched CDs properly, and that allow partially corrupted communications from faraway satellites to be corrected, could be generalized to stabilize quantum mechanical computations. We now have a family of quantum error-fighting techniques—including fault-tolerant quantum error correction, error-avoiding subspaces, and symmetrization. It seems that as long as we can continue to build precise quantum hardware, we can correct the physical errors.

Despite these advances, it is still not clear which technologies will be most suitable for implementing large-scale quantum computers. Those who experiment will first have to learn more about the various quantum mechanical systems, and then improve their ability to control and manipulate them. Theorists will have to suggest new ways to exploit these experimental capabilities. However, small-scale quantum computers, which will handle quantum information for other applications such as long-distance and quantum-key exchanges, are not far in the future.

What other kinds of applications can we anticipate? In the short term, applications will include ultra-precise quantum clocks. In the medium or long term, we can expect unconditional information security and large-scale quantum computers. These and other technologies will allow scientists to design and perform better experiments that will lead to new theories of physics, and continue the fruitful cycle of fundamental scientific discoveries and technological advancements.

Further Reading

Minds, Machines, and the Multiverse: The Quest for the Quantum Computer, Julian Brown (New York: Simon & Schuster, 2000).

The Code Book: The Science of Secrecy from Ancient Egypt to Quantum Cryptography, Simon Singh (New York: Doubleday, 1999).

"Quantum-enhanced information processing," by Michele Mosca, R. Jozsa, A. Steane, and A. Ekert, *Philosophical Transactions of the Royal Society: Series A*, Vol. 358, No. 1765 (January 15, 2000), pp. 261–79.

"From quantum code-making to quantum code-breaking," by Artur Ekert in *The Geometric Universe: Science, Geometry, and the Work of Roger Penrose* (Oxford: Oxford University Press, 1998). Available online at: http://xxx.lanl.gov/abs/quant-ph/9703035.

Born in London, England, in 1959, *Simon Lilly* and his family moved frequently when he was a child. He was living in the U.S. at the time of the Apollo moon landings, which he recalls as having a major impact on him and on his interest in astronomy. Educated at King's School Bruton in the U.K., Lilly later went to Sidney Sussex College at Cambridge University in 1977, to study natural sciences. He graduated with a first-class honours degree in theoretical physics three years later. Lilly subsequently studied for a Ph.D. at the University of Edinburgh under the supervision of Malcolm Longair, the Astronomer Royal for Scotland.

In 1984, Lilly moved back to the U.S. and worked as a postdoctoral fellow at Princeton University, then took up a faculty position the next year at the University of Hawaii. This gave him access to all the observatory facilities on Mauna Kea, including the Canadian national observatories. In 1990, he immigrated to Canada and accepted a position at the University of Toronto. In 1996, he spent a sabbatical year in Cambridge, U.K., as a Fellow of Clare Hall, and in 1997 was appointed a Fellow of the Canadian Institute for Advanced Research.

In October 2000, Lilly became the Director General of the Herzberg Institute of Astrophysics of the National Research Council. With a staff of one hundred and a budget of $18 million, the institute is responsible for operating all of Canada's national observatories, both in Canada and overseas.

Exploring the Universe in Space and Time

Simon J. Lilly

The telescope is a time machine.

The finite speed of light gives astronomers using powerful telescopes the remarkable ability to look at very distant objects and see them not as they are now, but as they were long ago. The light arriving from the distant corners of the universe shows us objects as they were when that light was first emitted. As a result, modern-day astronomers are uniquely privileged as Earth's only true time-travellers.

As astronomers, we have had the opportunity to travel through the farthest reaches of the Universe to witness its evolution, and to see that evolution, quite literally, laid out before our eyes. In the past, we have assumed that distant parts of the Universe are broadly similar to our own locale, but now, with increasingly powerful and sophisticated telescopes, we have evidence to back up that claim. On the largest scales, the Universe is, indeed, statistically homogeneous.

The quest to understand how the Universe has evolved, and how it came to have its present appearance, has been a driving goal for astronomers since Edwin Hubble's 1925 discovery of the immense scale of the Universe and the existence of large-scale galaxies beyond our own Milky Way. As we have sought to penetrate to ever greater distances, to peer ever further back in time, we have continually sought to build larger telescopes and to place more sensitive instruments on them. Our progress has been driven almost entirely by the technology of observation.

The nineties saw dramatic progress in this quest. We were able to detect and study galaxies similar to our own over an enormous span of cosmic time, ranging from the present epoch back to within about a billion years after the Big Bang (the cosmic explosion that essentially enabled these galaxies to form). While many teams of astronomers around the world have contributed to this progress, Canadians, who have access to some of the finest telescopes in the world, have played a very significant role.

I was fortunate to get into this field in its infancy, stumbling across the right thesis advisor at the right time. In 1980, Malcolm Longair, then the Astronomer Royal for Scotland at the University of Edinburgh, was discovering "radio-galaxies" at very high "redshifts." The redshift (z) is a measure of the speed at which a galaxy is moving away from us as a result of the expansion of the Universe. In fact, all distant galaxies show a redshift proportional to their distance from earth. More precisely, $1 + z$ is the factor by which the Universe has expanded in the time during which the light has been travelling to us. By 1980, Longair and his collaborators were finding radio-galaxies at redshifts of $z > 1$, which implied that the light had left these galaxies when the Universe was less than half its present age. These radio-galaxies were being found by virtue of their unusually strong radio emissions, which distinguished them from amongst the myriad faint galaxies that are visible on deep images of the sky.

It was not long before I also became involved. Longair set me up to study these radio-galaxies using the new United Kingdom Infrared Telescope (UKIRT) in Hawaii. Using the new technology of the infrared, we could look at the regular starlight that had been emitted in the visible waveband, but shifted to longer wavelengths by the large redshifts. Although at that time more distant objects were known—the enigmatic quasars—our radio-galaxies were the most distant objects known in which we could study recognizable starlight. Moving to Princeton University in New Jersey, and then to the University of Hawaii, I continued to study the radio-galaxies. It was becoming increasingly clear, however, that the radio-galaxies could not be regarded as typical of the general galaxy population. For instance, our own Milky Way galaxy is not a radio-galaxy now, and it almost certainly never has been one in the past. But finding more normal galaxies was hard: our own Galaxy, placed at a redshift distance of $z = 1$, would be very faint, ten times fainter than anyone had previously worked out, and only about one hundredth as bright as the "dark" night sky. However, together with Lennox Cowie, a colleague who was also in Hawaii at the time, I showed that we could successfully measure a redshift for such galaxies with long, eight-hour exposures on what was at the time the world's finest telescope, the Canada-France Hawaii Telescope (CFHT) on Mauna Kea in Hawaii. Our first paper in 1990 was based on just half a dozen objects—we could get one new redshift a night.

We were able to detect and study galaxies similar to our own over an enormous span of cosmic time, ranging from the present epoch back to within about a billion years after the Big Bang.

Technological innovation soon came to the rescue. David Crampton, at the National Research Council's Herzberg Institute of Astrophysics, and Paul Felenbok, at the Observatoire de Paris-Meudon, were already developing a novel spectrograph that would enable us to simultaneously observe 100 such galaxies. Very simple in concept, yet brilliantly engineered, their Multi-Object-Spectrograph (MOS) for the CFHT was going to be a world-beater.

Following my move to the University of Toronto in 1990, Crampton and I joined with the French astronomers Olivier Le Fèvre and François Hammer to undertake a systematic survey of these galaxies. After just twenty-two nights of observation, we had amassed over one thousand redshifts, and when we published the *Canada-France-Redshift Survey* in 1995, we could claim to have a survey ten times larger than any previous sample at this depth, and ten times deeper than any comparably large survey.

The *Canada-France-Redshift Survey* served to establish, unambiguously, that the population of "normal" galaxies had changed with cosmic time—finally settling a point that was being debated at the time. We found that star formation was more vigorous and widespread at earlier times in the life of the Universe, when it was about one half to one third of its present age. We also discovered that the galaxies appeared less regular in their structure and less developed than previously thought. These findings were not completely unexpected, given the immense "look-back" times involved. We also found evidence that the merging of galaxies was more common in the past. In short, we were clearly looking at a younger galaxy population. But along with our discoveries, there were some things of which we could find no evidence. We did not find the long-sought "proto-galaxies." We did not find any objects that looked like a Milky Way galaxy actually in the process of formation. As a result, we figured that the epoch of galaxy formation must lie still further back in time, and further out in space.

Technology and science quickly moved on. Within a year of publishing our survey, astronomer Chuck Steidel and his team at the California Institute of Technology successfully identified the first galaxies at even higher redshifts ($z = 3$). They used the giant Keck ten-metre telescope, the first of a new generation of telescopes and much larger than our own 3.6-metre CFHT. Due to the huge redshifts involved, it has not been easy to relate these new galaxies to the local Universe. Their nature is still debated. However, I was struck by how similar these objects were to our own *CFRS* sample at lower redshifts. I could see no compelling evidence for the proto-galaxies in their sample either. Were we missing something hidden? Perhaps.

The first clue that something was missing came from a 1996 discovery in space, when NASA's COBE satellite measured the integrated light from the extragalactic Universe (that is, the brightness of the entire Universe *outside* our Milky Way galaxy) over a broad

Along with our discoveries, there were some things of which we could find no evidence. We did not find the long-sought "proto-galaxies." We did not find any objects that looked like a Milky Way galaxy actually in the process of formation. As a result, we figured that the epoch of galaxy formation must lie still further back in time, and further out in space.

range of wavelengths. The result was interesting. There was as much light in the far-infrared end of the spectrum at a wavelength of 100~ m as there was in regular starlight at visible wavelengths around 1~ m. We knew that the far-infrared light was emitted by dust that had absorbed visible and ultraviolet starlight, and re-emitted the absorbed energy as a far-infrared emission at a temperature of about fifty kelvin. The Universe at visible and near-infrared wavelengths, which other researchers and I had been studying for so long, was clearly only half the story. We needed to understand the sources of the dust emission.

One of the virtues of the Canadian system of national observatories is that all Canadian researchers have access to a broad range of observational facilities, regardless of their institutional affiliation. Just as the question of the dust became pressing, our British partners at the U.K.-Netherlands-Canada James Clerk Maxwell Telescope in Hawaii introduced a revolutionary camera system. Known as SCUBA, it worked in the sub-millimetre waveband at 450 and 850~ m. For the first time, we could make deep images of the sub-millimetre sky and hope to detect the individual sources that were responsible for the integrated background detected by COBE. Steve Eales, Walter Gear (the developer of SCUBA), and I undertook a deep survey with SCUBA, successfully detecting a large number of discrete sources.

We are still studying these new objects to understand what they are, but it is already clear that they lie at high redshift distances, and they represent a class of rare, but very luminous, galaxies. It is likely that we are seeing a phase of galaxy evolution in which a large number of stars are produced in a very short amount of time—in regions that are heavily obscured by dust. There are similar obscured sources in the local Universe, which appear to be powered by enormous bursts of star-formation triggered by the merger of two galaxies. In the local Universe, such objects produce less than one third of a percent of the total luminous output of the Universe. At the earlier epochs probed by our SCUBA survey, however, such objects are producing more like thirty percent of the total luminous output of the Universe. This increase is likely related to the evidence for increased merging that we found earlier in our 1995 survey of red-shifts at optical wavelengths. However we look at it, this must be an important phenomenon in the cosmic scheme of things. These dust-enshrouded sources are emitting a good fraction of the luminous energy that has ever been produced in the Universe. We are confident that we have found an important piece in the puzzle of how galaxies form.

We seek to understand humanity's place in the Universe and to understand our origins. We seek to understand how we, and all that we see around us, came to be.

Even as we sketch out the history of galaxies over cosmic time, we know that our picture is still incomplete. We have not yet seen "first light" (the generation of stars that first lit up a dark Universe). We suspect that the first star clusters to form will be found at much higher redshifts ($z = 30$) than we have yet probed ($z < 6$). The Canadian Space Agency is a partner with NASA and the European Space Agency in developing the $1.4-billion Next Generation Space Telescope (NGST), a successor to the Hubble Space Telescope, due for launch in 2009. The NGST is being specifically designed to detect first light.

We are also far from understanding the physics that drive galaxy formation and evolution. Gas must be playing a key role in the assembly of galaxies. Although we know that stars are formed from gas, we know little about the gas in young galaxies. The Atacama Large Millimeter Array (ALMA), a $1-billion, multinational project involving the United States, Europe, Japan, and, we hope, Canada, will be an array of millimetre-wave telescopes at an altitude of five thousand metres in the Chajnantor region of the Chilean Andes.

The ALMA will finally allow us to study the physical conditions of gas in and around forming galaxies in the early Universe. The Canadian astronomical community has identified our participation in both the NGST and the ALMA as our highest priority for new investment in observational facilities over the next ten years.

We are on a voyage of exploration into the furthest reaches of the Universe, using very expensive telescopes to explore phenomena that are unlikely to materially affect us today. Why should we do this at all?

Our goal is simple. We seek to understand humanity's place in the Universe and to understand our origins. We seek to understand how we, and all that we see around us, came to be. Astronomers at the turn of the millennium have adopted the goal of gaining a causal understanding of the sequence of events that started with the Big Bang and ended with the emergence of DNA. This is an ambitious goal, possibly a presumptuous one, but one with a certain nobility and grandeur. It is appropriate that Canadians are playing a significant role in what has become a truly global endeavour.

Further Reading

The Origins of Our Universe: The Royal Institution Christmas Lectures for Young People 1990: *A Study of the Origin and Evolution of the Contents of Our Universe*, Malcolm S. Longair (Cambridge: Cambridge University Press, 1991).

Galaxies, T. Ferris (San Francisco: Sierra Club Books, 1980).

"Galaxies in the Young Universe," F.D. Machetto, and M. Dickinson, *Scientific American* (International Edition), Vol. 276, No. 5 (1997), pp. 66–73.

Dr. Mark Henkelman has always wanted to know what is inside things and how they work. When he was three, his mother caught him with the telephone dismantled and in pieces, asking, "How it work?" Things haven't really changed much since then.

After completing two degrees in theoretical physics, Dr. Henkelman moved into the field of biological physics and worked on electron microscopy of viruses—to see how they worked. He had his first faculty position at TRIUMF—Canada's national laboratory for particle and nuclear physics—investigating an experimental radiation treatment for cancer. This led to magnetic resonance imaging (MRI) and one of the first MRI machines in Canada at the University of Toronto and Princess Margaret Hospital. He has contributed extensively to the understanding of how MRI works and how to apply it to diseases like cancer, arthritis, and white matter disease.

Today, as a professor in the departments of medical biophysics and medical imaging at the University of Toronto, and a senior scientist in Imaging Research at Sunnybrook and Women's College Health Sciences Centre, Dr. Henkelman's research is focused on expanding the use of MRI technology further into the diagnosis and the management of therapy for cancer. In addition, he is a co-author of more than two hundred publications and author of more than three hundred abstracts, and has given numerous presentations on his research around the world.

Now, with a team of colleagues and students, Dr. Henkelman is developing medical imaging for mice—not just because it's challenging and fun, but because it helps to answer a very big question, "How does a mouse, and hence a human, really work?"

Does Your Mouse Need a CAT Scan? Medical Imaging of Mice

R. Mark Henkelman

If you are not feeling well, and your doctor cannot immediately diagnose the problem, then it is more than likely that you will be sent to a medical imaging department for a closer look at what ails you. There, you might be given an X-ray, or its three-dimensional version, a CAT scan. Maybe you will receive an ultrasound, an MRI, or a PET—typically used to study problems associated with the brain. All these scans look for anatomical abnormalities, such as a broken bone, a growing tumour, a congenital abnormality, a blocked coronary artery, or even multiple sclerosis plaques in the brain. They are all perfectly designed for humans.

But what if you were a mouse?

At Toronto's Hospital for Sick Children, a team is working to solve this problem. The team is building a Mouse Imaging Centre (MICe). The centre will provide for mice what a medical imaging centre in a regular hospital provides for people; that is, a spectrum of imaging instruments for viewing anatomy in the body, following the development of disease, and monitoring response to treatment.

MICe will provide X-ray, ultrasound, computer tomography, and magnetic resonance (MR) imaging, as well as optical microscopy, all adapted and optimized for imaging mice. The transition from human-scale medical imaging to the imaging of mice poses unique challenges and opportunities. Just because a mouse is smaller than a human doesn't mean that the imaging equipment will necessarily be smaller.

Before we get too carried away with images of ailing mice filling up hospital beds and sitting impatiently in emergency room waiting areas, however, we should take a few moments to explain why we need these mouse imaging facilities.

The cause behind many human diseases can be found in our genetic makeup. The genome is the long and complicated sequence of paired nucleotides, referred to as A, C, G, and T, that determines who, or what, an individual is. The genomes have been sequenced for the worm, the fruit fly, and for a rapidly growing list of other species. A draft sequence of the human genome was completed in 2001, and the mouse genome is following. We believe that the DNA sequence specifies normal development, likelihood of disease, and, most importantly, individualized response to treatment. Specific genes have already been identified as the causes of certain diseases, such as cystic fibrosis and sickle-cell anemia. The most common diseases result from the interaction of many genes, though. It is also becoming increasingly evident that a single gene can affect many different functions. This complex relationship between genes and the function of the individual (animal, plant, or human) is the major question facing biology and medicine in the twenty-first century.

Although we are ultimately interested in the relationships between human genes, human functions, and human disease, we know that these relationships will be charted primarily in the mouse. That is because, for any gene that can be identified in the human genome, there is a ninety-five percent probability that an equivalent gene can be found in a mouse. Those genes that have been identified in mice have functions very similar to those in humans.

In addition, the reproductive cycle of the mouse is comparatively short, enabling rapid expression of particular genomes over numerous generations. The sequence of the mouse genome will provide the underlying detailed genetic sequencing information for interpreting changes in function. As well, molecular procedures for modifying genetic information have been worked out for the mouse, and the procedure to grow transgenic (modified genes) and knockout (deleted genes) mice is becoming routine. As a result, the mouse is an ideal mammalian model system, in which the complex relationship between genes and functions can be worked out in detail, and from which the story can be adapted to the human and to human diseases.

The research starts with an inbred strain of mice, such as the C57-Black, a standard laboratory research mouse for which the particular genome sequence will become available. After changes are made in the DNA sequence, either randomly using chemical mutagenesis or in a targeted manner, a new mouse can be grown by inserting the modified DNA into an egg cell from which the nucleus has been removed. How will we know if this new mouse with an altered DNA is developing abnormally or is getting sick? In exactly the same ways that we find out whether a person is getting sick. We do it by conducting a physical examination, and monitoring its weight, temperature, heart rate, and blood and urine chemistry. Most importantly, we do it by medical imaging.

For any gene that can be identified in the human genome, there is a ninety-five percent probability that an equivalent gene can be found in a mouse.

The X-ray used for the imaging of mice is a scaled-down version of the human X-ray. A micro-focus X-ray source shines through the mouse and is detected, not with film, but with a direct digital detector adapted from a charged-coupled device camera. Computed tomography can be realized by rotating the X-ray around the mouse and feeding the digital images directly into a computer. One limitation to X-ray imaging of mice is the dose of radiation. To achieve human-type image resolution from a mouse, the image-element volume needs to be one thousandth of that used in human imaging, requiring a radiation dose that is far greater than that used for human X-ray or CT. Such a radiation dose traditionally restricts CT to imaging dead mice, and to evaluating bone structure and blood vessel patterns.

Ultrasound is easier in mice because the sound does not need to penetrate into deep abdominal regions, as it does in human imaging. This allows for higher-frequency and higher-resolution imaging (down to one twentieth of a millimetre). The real-time advantage of ultrasound is preserved, making it ideal to study the heart and fetal development.

Magnetic resonance microscopy in the mouse can achieve very high resolution, exceeding one twentieth of a millimetre in all directions. This is made possible by increasing the strength of the imager's magnetic field to five times greater than is customary in human MR. The system is designed to image sixteen mice at the same time, in order to increase the total number of mice that can be imaged in a year. These factors mean that the mouse-imaging magnet will be even bigger than it is in a human MR imager. It will be approximately fifty tonnes, and the size of a small truck.

Optical microscopic imaging has little use in human medical imaging, but because of its exquisite resolution, it is ideal for generating three-dimensional images of early mouse fetuses, with cellular resolution. When used with specific dyes and stains, optical imaging can identify specific gene expression.

The Mouse Imaging Centre at The Hospital for Sick Children will work in concert with a large research program called the Canadian Models of Human Disease (CMHD), which is headed by Canada's leading developmental biologist, Dr. Janet Rossant. In a very broad search to find out which genes are related to dysfunctions that look like human diseases, the CMHD program will make chemically induced, random mutations in the mouse genome, and produce many thousands of individual, genetically different mice per year. As part of the screening program to look for diseases, MICe is designed to image these large numbers of distinct-genome mice. Thus, the imaging equipment is designed either to image rapidly, or to image many mice in parallel.

Genetically altered mice may be studied for heart or brain abnormalities, and then, somewhere else in the world, somebody could study these same images, looking for pancreatic disease. A mouse that had perfectly normal heart development could be exactly what is needed for a pancreatic disease model.

This large-throughput anatomical survey is only the first step for MICe. After a disease model has been found, imaging will again be used to characterize the progression of disease over time in the same individual mouse. For diseases arising from developmental abnormalities, it is the progressive deviation from normal development that is most informative. For adult diseases such as cancer and neuronal degeneration, the natural history of the disease is a major part of the disease description. The ability to use imaging to study individual mice over time is invaluable in characterizing these diseases. Even more importantly, using imaging for the sequential evaluation of disease progression provides the means to monitor response to treatments. As a result, MICe will also be a major resource for researchers working on experimental treatments.

Medical imaging is playing an increasingly major role in the guidance of treatments such as image-guided surgery and gene therapy. The same will be true of MICe, where image guidance will be used to reintroduce genetic material to specific locations during the course of development. These studies are often essential for a definitive demonstration of the effects of changed or deleted genes.

MICe has been conceived as a nationwide resource. Investigators across Canada with genetic models of disease will be able to send mice to the centre for image characterization. Conversely, the image data from MICe will be made available to researchers around the world. This represents a monumental task, since the three-dimensional image data from a single mouse is about the size of a small encyclopedia, and will barely fit on a single CD. Fortunately, the expanding speed and capabilities of the Internet will provide the necessary capacity for the wide distribution of such large amounts of data. It is very important that there be widespread access to this imaging data. Genetically altered mice may be studied for heart or brain abnormalities, and then, somewhere else in the world, somebody could study these same images, looking for pancreatic disease. A mouse that had perfectly normal heart development could be exactly what is needed for a pancreatic disease model. Thus, the careful, systematic, and accessible archiving of this imaging data, along with the specific genomes and phenotyping, will constitute an invaluable resource for future data mining—to find the relationship among genes, structure, function, and disease. Such an endeavour will inevitably involve close collaboration between academe and industry, medical imaging and biology, computer and instrumental engineering, and pharmaceutical companies and molecular biologists. MICe is a nexus to stimulate these interactions.

There is a long road ahead that leads to understanding the genetic bases of disease. Genome sequences are just the beginning. Gene products—the proteins—must be identified and their structure, interactions, and functions determined. The regulation of gene expression—which genes are active and which are not—needs to be understood over the course of development, as does its variation throughout the different tissues of the body. It is a tremendous task, and it is not yet clear how all the information will be put together into coherent stories. It is clear, however, that mouse imaging has a critical role to play.

At some point in your future, medical imaging will discover that a disease can be treated based on your specific genetic makeup. When that happens, you can be sure that a mouse was imaged before you.

Further Reading

"Scanners Get a Fix on Lab Animals," R.F. Service, *Science*, Vol. 286 (1999), pp. 2261–63.

"Genomics: Journey to the Center of Biology," E.S. Lander and R.A. Weinberg, *Science*, Vol. 287 (2000), pp. 1777–82.

http://www.cmhd.mshri.on.ca

http://mouseimaging.bioinfo.sickkids.on.ca

Farmed seafood

Dr. Joseph Brown's research program at the Ocean Sciences Centre of Memorial University of Newfoundland is composed of both fundamental and applied research. His fundamental area of interest is applying theories of behavioural ecology to studies of larval and juvenile fish. His applied area is aquaculture, where his research laboratory focuses on developing new species for cold-water aquaculture and on improving culture techniques for various cold-water species.

He has been at the Ocean Sciences Centre since 1985, and is currently a theme leader within AquaNet, a recently funded addition to the Networks of Centres of Excellence, focusing on aquaculture research. He is past president of the Aquaculture Association of Canada, and he served on its board of directors from 1991 to 1998. As well, he is past chair of the APICS Aquaculture Committee, and is currently the chair of the M.Sc. (aquaculture) program at Memorial University.

Dr. Brown received his B.Sc. from St. Francis Xavier University, his M.Sc. from Memorial University of Newfoundland, and his Ph.D. from Queen's University.

From Fishing to Farming: The Domestication of New Species for Aquaculture

Joseph A. Brown

Two centuries ago, in adapting to our land habitat, we went from a hunter-gatherer society to a farming society. In North America, we are now slowly going through a new and similar transition with our fishing culture. In many rural areas of Canada, traditional fishing is being replaced by fish farming, or "aquaculture."

Aquaculture is a new industry in North America, and one which depends on a natural resource. It is an industry coming of age at a time when environmental awareness and concern is growing. Aquaculture is the farming of aquatic organisms, including fish, molluscs, crustaceans, and aquatic plants. It involves intervention to enhance production with some form of individual or corporate ownership of the organisms produced. Aquaculture is not a "fishery," and with the well-documented decline in many of the wild fish stocks, interest in aquaculture to produce seafood for consumption has heightened.

Over the past twenty years, aquaculture has been a growth industry in Canada. Farmed seafood now makes

Over the past twenty years, aquaculture has been a growth industry in Canada. Farmed seafood now makes up twenty-nine percent of the fish harvested for human consumption.

up twenty-nine percent of the fish harvested for human consumption. Currently, over eighty percent of the value of Canadian aquaculture is based on the production of salmonid fishes (salmon, trout, charr), and there is interest in diversifying this base. As well, many aquaculture salmonid businesses have made money, and they are now interested in diversifying their product lines. These are two reasons for domesticating new candidate species for aquaculture. Other reasons include the desire to develop new products for a growing market, and the need to provide job opportunities in those rural areas of Canada hit hard by the recent decline in the traditional capture fishery.

Among the new candidate species are two native finfish species found in Atlantic Canada that are just starting to be cultured on a commercial scale. But first, we should keep in mind that not all aquatic organisms are suited to aquaculture. When considering which species are suitable, we must consider market value, production costs, and the quality of the product. While these are indeed valid considerations, the biological perspective must also be carefully considered—especially in the Atlantic region, where cold water (at temperatures less than zero degrees Celsius) persists for a significant portion of the year. This means that species chosen for aquaculture development can only present a reasonable market and production outlook if they are suited to the geography and climate in which they are grown. Biological and commercial considerations are linked because each affects production costs and output.

The reasons and criteria for developing new finfish species are, therefore, straightforward enough. Unfortunately, the time from initiation of research to commercialization of the species can be extensive, and time frames of fifteen to twenty years to prepare a species for domestication are not uncommon. In Europe, for example, research on turbot, sea bass, and sea bream began in the early seventies, but it wasn't until the mid-nineties that large numbers of juveniles were produced and ready for stocking in sea cages in

the Mediterranean. It takes so long because we often attempt to domesticate wild fish for which we do not have a lot of scientific information. As a result, we have to take the time to carefully develop basic production protocols. Once protocols are established, pilot-scale commercial facilities can be set up, with research continuing in order to fine-tune production.

Two finfish species of particular interest to Canada are the Atlantic halibut and Atlantic cod. Both species are being closely studied in Eastern Canada and elsewhere in the world, and a good deal of research has been done to establish their commercial potential.

The Atlantic halibut is probably the best known of the new finfish candidate species, and has been the focus of much research in Scotland and Norway since the early eighties. Successful metamorphosis (the normal anatomical and physiological change that occurs very early in the life cycle) and successful juvenile production were first accomplished in North America in 1993, by researchers at the Department of Fisheries and Oceans' Biological Research Station in St. Andrews, New Brunswick. The Atlantic halibut is an extremely difficult species to work with because of its unique early-life development, and because of the complications associated with its metamorphosis. On the other hand, it is a good example of a species with excellent market value because it is in low commercial supply and is characterized by good survival after it completes metamorphosis. It is considered a particularly attractive candidate for marine aquaculture because it is a fast grower, has high yield, and becomes sexually mature only after it attains a large size.

Not all aquatic organisms are suited to aquaculture. When considering which species are suitable, we must consider market value, production costs, and the quality of the product.

In halibut aquaculture, there are five stages—from spawning to the juvenile stage. Each stage presents its own problems and challenges. The broodstock (mature, spawning fish) represent the most important phase of any successful aquaculture venture, since the production of high-quality gametes (eggs and sperm) is critical

Spawning is not a simple matter. Halibut are batch spawners, and eggs must be collected from the female within six hours of ovulation. A female can spawn between five and fifteen batches of eggs during a season, and can yield a volume equal to forty percent of its body weight during a spawning period.

for good larval survival and juvenile production. Adult halibut are large and generally reside in deep water. Thus, halibut broodstock are maintained in the dark in holding tanks varying in diameter from five to eleven metres.

Spawning is not a simple matter. Halibut are batch spawners, and eggs must be collected from the female within six hours of ovulation. A female can spawn between five and fifteen batches of eggs during a season, and can yield a volume equal to forty percent of its body weight during a spawning period. With recent advances in the field of reproductive physiology that permit manipulation of the light cycle, spawning can be controlled. This means that up to three separate broodstocks can be manipulated to spawn at three different times during the year.

The larvae released at hatching are poorly developed. They are fairly big (six or seven millimetres in length) and possess no functional eyes or mouth, but have a very large yolk sac, which is their major source of energy. Depending on temperature, the duration of this stage extends between thirty and sixty days, until the eyes and mouth become functional. During this time, they are kept in the dark, at high salinity, and at six degrees Celsius. High mortality is common at this stage, but that improves once the larvae are able to capture prey. They are then transferred to the first feeding areas, where their diets consist of different combinations of zooplankton and rotifers.

After the larvae have metamorphosed and taken on their recognizable flat shape, weaning to artificial food begins. The growth and survival of juvenile halibut are readily achieved under proper aquaculture conditions. Two- to four-year-old halibut, kept in water between four and thirteen degrees Celsius, can triple their weight over a sixteen-month period.

In the spring of 1993, the first farmed halibut were sold on the market in Norway. This marked the start of cultured halibut commercialization. That first year, about five tonnes were sold. Current global production is over one hundred tonnes. The production of juveniles has occurred quite recently at three private hatcheries in the Maritimes. Within a couple of years, farmed Atlantic halibut from these operations will be on the market.

Cod fishing has been practised in Newfoundland's coastal waters since the fifteenth century, and until the recent moratorium, cod was the prime species captured there.

Cod fishing has been practised in Newfoundland's coastal waters since the fifteenth century, and until the recent moratorium, cod was the prime species captured there. As a result, the inclusion of Atlantic cod as an aquaculture species seemed unlikely. However, with the alarming decline in cod stocks and the long-term moratorium imposed on the fishery in Atlantic Canada, the farming of this species no longer seems unrealistic. In fact, "cod culture" is not a new venture. In the 1890s, Adolph Nielsen, a Norwegian fisheries inspector, was hired by the Fisheries Commission of the Newfoundland government to establish and manage a cod hatchery to "restock our exhausted bays." A little over a century later, the situation in Newfoundland shows a remarkable similarity. This time, however, the situation has resulted in the closing of most of the cod fishery on the province's east coast.

An element that contributed to a successful new method of cod farming in Newfoundland and Labrador after 1986 was the abundance of live cod caught in what is called the "trap fishery." Trap fishing is a method unique to Labrador and Newfoundland. In other countries, cod are principally caught using large pots, otter trawl, or long line. Cod captured in traps are transported live to marine, near-shore sites where they are then grown for market. Until the moratorium, cod were "grown out" in pens, similar to

With the declining availability of cod, there has been a concerted marketing effort to present it as an upscale menu item. The result has been a surge in interest in the prospects of farming this species.

those used on salmon and trout farms, with cage sizes varying between six hundred and fifteen hundred cubic metres. Once the fish became acclimatized to the pens and begin to feed, they double their weight in three to four months. If fish weighing more than one kilogram are stocked in pens at appropriate densities in spring and fed adequately, and if water temperatures are not too high, then the bulk of the stock will be harvested between November and December of the same year. If fish do not achieve the desired harvest size of two kilograms or more, they can be successfully "wintered" with minimum attention in areas with land-fast ice cover—until temperatures rise above zero degrees Celsius the following spring. Temperature and initial size are the two most important factors in determining the rate of growth in fish. Northern cod display optimum survival rates, food conversion, and growth at temperatures ranging between eight and twelve degrees Celsius.

Cod are harvested from farms and delivered live to fish plants. This ensures fillets of highest quality, and the superior quality of byproducts such as liver, stomachs, and gonads. With the declining availability of cod, there has been a concerted marketing effort to present it as an upscale menu item. The result has been a surge in interest in the prospects of farming this species.

With the onset of the moratorium in the early nineties, the capture of cod from traps was prohibited. That's when the farming of cod from egg to market was initiated. Raising cod from eggs had met with mixed success up until the mid-eighties. Collecting fertilized cod eggs from natural spawnings is less difficult than it is with halibut. Unlike halibut, cod yolk sac larvae are small at hatch (about five millimetres in circumference), and begin feeding on plankton much sooner (three to four days after hatch at ten degrees Celsius). Moreover, in contrast to halibut, in which meta-

morphosis is abrupt and results in quick morphological, physiological, behavioural, and habitat changes, metamorphosis in cod larvae is much more gradual. Once larvae attain a length of two and a half millimetres, they are considered juveniles.

Producing cod juveniles in hatcheries is now seen to have a number of advantages over the "catch-and-grow" technique. One advantage is that all cod placed in cages will have been weaned onto dry, formulated feed. This makes the logistics of feeding much more attractive when compared to holding large amounts of "trash" fish, such as male capelin and herring for feeding. Another advantage, given the year-round supply of commercial feed, is that this type of cod can be put in the water at various times, and can be harvested over a longer period than the catch-and-grow cod. Also, with the development of broodstocks that will spawn two to three times per year, the availability of juveniles is not a problem.

From a commercial perspective, the production of cod for sea-cage culture is a very new enterprise, and much research and development is required to demonstrate the commercial value of this venture. This is now happening, and shortly, farmed cod will be available in supermarket fish sections and take its place among the growing number of farmed aquatic organisms that will make their appearance over the next ten years.

Production of seafood from traditional fisheries has peaked and will not increase. Although the types of species coming from the traditional fishery will change, it is extremely unlikely that the overall amount of seafood produced from the fishery will grow. Farmed products, on the other hand, which already make up over twenty-five percent of the seafood consumed globally, will increase in percentage as more marine and freshwater species are domesticated. This is a significant change, and one that will have different benefits for various parts of the Atlantic region. Given the pressures on the planet's production systems, this shift from capture fishery to fish farming should prove beneficial and sustainable. The reasons for optimism regarding the sustainability of aquaculture lie in the technological developments that will enable us to achieve efficiencies in production, while at the same time lessening the effect of these activities on the environment. As well, because aquaculture is easy to observe, we can monitor and adjust activities to make it more environmentally friendly and sustainable.

It is somewhat ironic that it was developments and advancements in fishing technology that enabled the traditional fishing industry to become too efficient for its own good. As technology develops further, will the same thing happen again? If used well, aquaculture can ensure a sustainable and varied supply of seafood for many years to come.

Further Reading

Special Bulletin on Marine Finfish Culture, edited by A. Boghen (1995), pp. 341–62.

Special Bulletin on Marine Finfish Culture, edited by J.A. Brown, Vol. 98, No. 1 (1998), pp. 1–40.

"Larviculture of Mediterranean marine fish species: Current status and future trends," J.W. Sweetman, *Journal of the World Aquaculture Society*, Vol. 23 (1992), pp. 330–37.

In his position as Assistant Chief Statistician, Analysis and Development, at Statistics Canada, *Dr. Michael C. Wolfson* is responsible for analytical activities in general, for health statistics, and for specific analytical and modelling programs.

Before joining Statistics Canada, Dr. Wolfson held a variety of positions in central agencies—including the Treasury Board Secretariat, Department of Finance, Privy Council Office, House of Commons, and Deputy Prime Minister's Office—with responsibilities in the areas of program review and evaluation, tax policy, and pension policy.

Dr. Wolfson has been a Fellow of the Canadian Institute for Advanced Research since 1988. His recent research interests include income distribution, tax/transfer and pension policy analysis, microsimulation approaches to socio-economic accounting and to evolutionary economic theory, design of health information systems, and analysis of the determinants of health. Dr. Wolfson received his B.Sc. in computer science and economics from the University of Toronto in 1971, and his Ph.D. in economics from Cambridge University in 1977.

Social Proprioceptors—Understanding the Health of Canadians

Michael C. Wolfson

How long do Canadians survive after being diagnosed with cancer? What difference does being physically fit make to your chances of having a heart attack? How much longer do heart attack patients live after bypass surgery?

Although these questions are simple enough, the answers are anything but easy. Before we can even begin to formulate thoughtful and meaningful answers, we need access to one important thing: systematic, high-quality data. It's only with such data in hand that we can get a clear picture of the state of health of Canadians and of their health care system. Now, after years of data drought, the details are beginning to emerge.

Before we launch into our examination of data and health, an explanation of this essay's title and the word "proprioception" is in order. Our nervous system is composed of different kinds of cells. Motor neurons communicate signals to our muscles. Sensory neurons bring signals about external stimuli back to the brain.

Proprioceptors, another group of neurons, continually update the brain on our body's position and orientation. The country's statistical system is a form of *social* proprioceptor, continually providing us with vital information on where we are and where we're heading.

Unfortunately, in the area of health care, our social proprioceptors are still not fully developed. But there is hope. The deficiency has been recognized at the highest federal and provincial levels, and plans are already underway to begin a process of systematic data collection, and of reporting to Canadians on the status of their health and the functioning of their health care system.

This development builds on more than a decade of dedicated work across Canada involving researchers and thoughtful government officials, among others. My personal involvement began in the late 1980s at Statistics Canada, when I was asked to write a "think piece" on the state of our health statistics program. My first conclusion was that the data we were collecting and publishing were a hodgepodge—an incoherent mixture of death rates, hospital bed days, doctors per capita, health care costs as a percentage of GDP, and so on. Each bit might be interesting, but there was no framework to hold the pieces together—nothing comparable to the System of National Accounts that gives us regular proprioceptive feedback on the state of the nation's economy.

My second observation was the near-complete absence of any information on how healthy Canadians are. How is it possible to judge the activities of the health care system without any bottom line, without any systematic measure of the population's health and of how the health of individuals is being influenced by the billions of dollars spent on cures and care?

The obvious way to respond is simply to follow individuals over time to build up evidence on what works, gathering information about their health both before and after various interventions like surgery. But things are never as simple as this. Even though this idea is at least a century old (a Boston physician named A.E. Codman is credited with first developing the approach to keep track of patients), it's not yet standard practice for managers of Canada's health system.

Statistics Canada took a major step forward with the launch of the National Population Health Survey in 1994. Conducted every two years, the survey fills a major gap by providing a series of snapshots that show just how healthy Canadians are. The information can be broken down by province, sex, age group, aspect of health status (for example, chronic conditions, disability, ability to carry on usual social roles), and socio-economic status.

Also, the survey incorporates two major innovations. First, it is longitudinal. This means that respondents in the first wave of the survey are followed up on and contacted every two years. As a result, for the first time on a nation-wide basis, we can begin to answer one of our opening questions: Does keeping fit reduce your chance of having a heart attack? The answer is yes. The survey also shows that the protective benefits of being fit apply even when you are overweight.

The country's statistical system is a form of social proprioceptor, continually providing us with vital information on where we are and where we're heading.

The other major innovation was to ask survey respondents for permission to allow Statistics Canada access to the data in their provincial health care records. The vast majority agreed. Armed with this important link, we are now beginning to assemble data to address such seemingly simple questions as: What have been the benefits of the flu vaccine? Which groups have not been getting vaccinated? What differences has it made?

If these two major innovations are so simple and have such potential power in terms of their health benefits, why hasn't anyone acted on them before? There are several reasons for this, one of which is simply inertia. It takes time for discussion and consensus-building to shift the focus of a large, complex system with many diverse players. Another is lack of imagination—the failure to think through the huge potential benefits of statistical quality control. This has been common in the manufacturing sector for decades, but has yet to be applied to health care in the same way. A third reason is the failure of the health care system to use the computers, networks, and database software that have been so effective in other areas, such as the airline and banking industries.

Two other reasons raise legitimate concerns. One is individual privacy. Assembling detailed longitudinal histories of individuals' health care encounters with their doctors and other providers—combined with questions on such issues as their health status, education, risk factors, and labour market experiences—is a de facto invasion of privacy. On the other hand, with names and addresses stripped off, the data is used to produce strictly statistical results. At Statistics Canada, such record linkages are approved on a case-by-case basis when there is a clear expectation of an important public benefit. Although privacy concerns are legitimate, they are judiciously balanced against the potential benefits of gaining a better understanding of what makes Canadians healthy.

The other reason is slow evolution in the culture of the medical profession. Doctors have not been quick to seek this kind of information. One might think they would be enthusiastic about information systems that would help them be more effective and efficient in their practices. However, computers and databases have not figured prominently in medical school training (though this is changing). Also, doctors' remuneration and their daily routine of treating patients are not conducive to entering data or getting computer-based suggestions on diagnosis or treatments.

Medical practitioners might also feel concerned that these kinds of computerized systems would put them on the receiving end of evaluations of their "practice patterns," perhaps pointing out things they are doing that are inappropriate, and not supported by the evidence on what works. However, it is just as likely that any problems or inefficiencies that become visible are systemic, due to problems of coordination among the many different providers across the "continuum of care" with whom a seriously ill individual typically comes in contact.

The longitudinal health information that will be generated for representative samples of individuals has implications well beyond the health care system. For example, we know that poor people are generally in poorer health. For over a decade, I have

It is not simply the case that once we move above some poverty threshold, everyone is more or less equally healthy. Rather, every step up the socio-economic ladder seems to confer an incremental increase in health status.

had the privilege of being a Fellow of the Canadian Institute for Advanced Research's Program in Population Health. This program has made seminal contributions to our understanding of the broader determinants of health. One of the most pervasive findings is not only that poor people are less healthy, but that there is a gradient phenomenon. It is not simply the case that once we move above some poverty threshold, everyone is more or less equally healthy. Rather, every step up the socio-economic ladder seems to confer an incremental increase in health status.

Longitudinal data are not only central to understanding the operation and effectiveness of Canada's health care system, they are also fundamental to advancing basic knowledge about the determinants of health, such as the role of socio-economic factors. The socio-economic gradient now seems to be part of conventional wisdom, at least in Canadian health policy circles. It wasn't a decade ago, though. At that time, there was a great divide between the kinds of evidence considered in health debates. On one hand, clinical medicine had built up methods of randomized clinical trials, with very careful measures of endpoints and precise statistical methods. From this kind of research, for example, there was unequivocal evidence that certain drugs were effective in lowering an individual's cholesterol levels.

On the other hand, the evidence in Canada for the socio-economic gradient was much weaker. Only the odd survey showed a correlation between how healthy respondents said they felt and their level of income or education. But correlation is not causation, and there were competing explanations for this observation. It could simply have been that individuals got sick and their incomes fell as a result.

To shed light on this issue, we obtained data (under the protection of the *Statistics Act*) on over half a million male Canada Pension Plan contributors and retirees. We then examined the relationship between these individuals' earnings when they were age forty-five to sixty-four, and their survival from age sixty-five to seventy-four. The results were dramatic and incontrovertible. We observed a clear gradient between pre-retirement earnings and post-retirement survival. Moreover, for the majority of cases, the causality had to run from socio-economic status to mortality, and not the other way around. And how large is this gradient? If the eighty percent of men with the lowest earnings were somehow able to achieve survival rates in the top fifth, their increase in life expectancy would be roughly one year—the same as if cancer had been completely eliminated as a cause of death.

Establishing the profound importance of socio-economic factors as determinants of health has still been much easier than figuring out the causal pathways by which they exert their influence. Without this knowledge, it's foolhardy to propose what to do. For instance, researchers mapping the human genome know that mapping is only the first step in developing effective new biochemical interventions. The next step is figuring out the sequences of proteins produced by the various genes.

It's the same for those of us who gather data. We now have to move on to the hard work of gathering appropriate data and developing the analytical tools to link together the information that researchers are currently uncovering. Richly detailed longitudinal surveys are required in order to assess the existence and quantitative strength of relationships among physical activity, diet, education, smoking, work environment, health care, immune system resilience, and health.

Piles of data and research results, interesting as they are, will not be enough. Complex simulation models need to be developed that can bring together all the disparate pieces of research and turn them into an analytical tool that can help us understand the complex web of significant interactions. It is for this reason that Statistics Canada has been developing the POpulation HEalth Model (POHEM).

POHEM started out as nothing more than a "thought experiment," part of the think piece I wrote over a decade ago. It is now operational and is being used to inform health policy in areas like cancer screening and assessing the likely benefits of approving a new drug. It is also being used in more basic research, for example, to estimate how much of the socio-economic gradient in health is attributable to the observed gradient in smoking, and how much remains to be explained by other factors. POHEM is also being used to establish a framework of population health indicators. These indicators will measure the overall burden of illness among Canadians in a way that allows us to connect changes in their health status to health care practices and other factors.

With new health surveys, the development of person-oriented health information systems, and the construction of sophisticated simulation models at Statistics Canada, we are finally beginning to build the social proprioceptors that Canada needs. This is also taking place more broadly across the country, with recently increased funding for both health research and health information systems. These social proprioceptors will allow us to understand and manage our health interventions in ways that will improve and promote health far more effectively. In these areas, Canada is widely regarded as a world leader. When I've described some of these Canadian initiatives to colleagues in the United States and around the world, they invariably respond, "If only we could do such things."

If the eighty percent of men with the lowest earnings were somehow able to achieve survival rates in the top fifth, their increase in life expectancy would be roughly one year—the same as if cancer had been completely eliminated as a cause of death.

Of course, these developments represent the dedication of a growing community of individuals who understand the broader implications of investing in the systematic measurement of population health status. They are a diverse group, including politicians, senior public servants, and university-based researchers, as well as managers and providers caught up in the day-to-day work of providing health care. It has been very rewarding to play a role in these endeavours. Today, I remain as excited as ever at the prospects these new avenues of information development, analysis, and research bring to our ability to understand, and improve, the health of Canadians.

Further Reading

Why are Some People Healthy and Others Not? The Determinants of Health of Populations, edited by Robert G. Evans, Morris L. Barer, and Theodore R. Marmor (New York: Aldine de Gruyter, 1994).

"The Health Template." Software developed as part of the 1990 report of the Task Force on Health Information. Available online at: www.statcan.ca/english/spsd/helthtem.htm.

Canada Health Infoway: Paths to Better Health: Final Report of the Advisory Council on Health Infostructure (Ottawa: Health Canada, 1999). Available online at: www.hc-sc.gc.ca/ohih-bsi/pubs/1999_pathsvoies/info_e.html.

"First Ministers Accord on Health," news release (September 2000). Available online at: www.scics.gc.ca/cinfo00/800038004_e.html.

Dr. Fred J. Longstaffe's research and teaching centre around the stable isotope chemistry and mineralogy of natural systems. He earned his B.Sc. in geology (with highest honours) and the Medal for Geology from the University of Windsor. He also attended McMaster University on a 1967 NSERC science scholarship, graduating with a Ph.D. in geology in 1978. Currently, he is the director of the Division of Earth, Ocean and Atmospheric Sciences, Academy of Science, at the Royal Society of Canada.

Awarded an Izaak Walton Killam Memorial Postdoctoral Fellowship at the University of Alberta, Dr. Longstaffe subsequently joined the university's faculty. In 1987, he was made a full professor of geology at the University of Western Ontario. Six years later, he became the first chair of Western's Department of Earth Sciences, and in 1999 was appointed Dean of Science. In addition, he was chosen as Western's Research Professor in Science in 1989 and its Florence Bucke Science Prize winner in 1990.

Dr. Longstaffe has served as president of the Geological Association of Canada. He has also been a council member for the Clay Minerals Society; the Association of Professional Engineers, Geologists and Geophysicists of Alberta; and the Mineralogical Association of Canada. He has served on the editorial boards of several scientific journals, and is the author or co-author of about two hundred scientific publications. He has been a member or chair of four NSERC research grant selection committees, and was a member of NSERC's Committee on Research Grants, where he was group chair for earth and environmental sciences. From 1993 to 1996, he represented Canada on the Advisory Panel for Advanced Study Institutes of NATO, which he chaired.

In 1993, Dr. Longstaffe was awarded the Past-Presidents' Medal for Research and was named Distinguished Fellow by the Geological Association of Canada. In 1997, he was elected to the Royal Society of Canada, and received the Mineralogical Association of Canada's Past-Presidents' Medal for research in 1998.

He is married to Linda (née McLean) and has four children—Jeffrey, James, Meghan, and Matthew. They live on the McLean family farm near Strathroy, Ontario.

Life as a Stable Isotopist: Romancing the Stones and More

Fred J. Longstaffe

A long time ago, while on a collision course with law school that had been charted since my youth, I happened upon a B.Sc. program in geology. Along with the opportunity to learn about rocks, minerals, oil, gas, and ancient life, came a chance to leave my part-time employment as a grocery clerk and instead traverse the Canadian northlands. This meant flying about in float planes, living under canvas in the wilderness, and, of course, looking for gold. As my interest in law waned, my ambition to explore the earth sciences grew. I never found gold, but some three decades later, I am still on that quest.

Quite by chance, my travels over the years have drawn me into the astonishing world of stable isotopes. And since then, one opportunity after another has followed. Stable isotopes have allowed me to examine "big picture" problems, such as climate change and global interactions among the atmosphere, biosphere, hydrosphere, and lithosphere. They have permitted me to investigate more immediate economic problems, such as

the efficacy of the steam stimulation process for in situ recovery of oil from Alberta's oil sands. Stable isotopes have even been my entrée into fascinating aspects of anthropology, such as the domestication of deer and dogs by the ancient Maya and its cultural significance. By understanding a few simple principles about the different varieties of atoms known as stable isotopes, and by discovering a little more about how they are distributed in nature, a powerful tool to investigate our natural world became accessible to me.

So what exactly are these stable isotopes? First of all, isotopes of an element are atoms. The nuclei of isotopes contain the same number of protons (Z) as the nuclei of the common element, but a different number of neutrons. The word "isotope" is derived from the Greek for "equal places." Isotopes of any given element occupy the same place in the well-known periodic table of the elements. For example, all isotopes of carbon contain six protons, and have an atomic number of six. This defines their collective place in the periodic table. Radioactive isotopes decay at a known rate, and can be used to date substances. Most of us are familiar with the concept of "carbon dating," in which the radioactive isotope of carbon (carbon-14) is used. Stable isotopes, by comparison, do not undergo radioactive decay. They are, as we say, *stable*.

How is it that these stable isotopes have had such a positive impact on my life? A key feature of the research I have been involved in is the opportunity, indeed the absolute necessity, to collaborate with experts from across the sciences, social sciences, engineering, and medicine. The greatest rewards have been the sheer joy of learning and the variety of discovery. When starting down this path, I could never have imagined the number of things I would learn. One year, I would be learning how to "read" the stable isotope record held by minerals and organic matter in old soils to discern climate change. In another year, I would learn how to track ancient movement of fluids deep in sedimentary basins, and to understand their relationship to the accumulation of metal and hydrocarbon deposits. Then I would learn how to deduce that an ancient mummy was a vegetarian, consumed a lot of maize, and

died because of famine and dehydration (as opposed to warfare or sacrifice). I learned all this through stable carbon and nitrogen isotopic analyses of tooth, bone, hair, and fingernail fragments. In the stable isotope business, you are what you eat.

The light elements, such as hydrogen (Z = 1), carbon (Z = 6), nitrogen (Z = 7), oxygen (Z = 8), silicon (Z = 14), and sulphur (Z = 16), which are so important in organic and inorganic interactions in natural systems, are each composed of at least two stable isotopes. Carbon, for example, is composed mostly of carbon-12 (six protons and six neutrons), but also contains small amounts of carbon-13 (six protons and seven neutrons), in addition to a trace of radioactive carbon-14 in some materials. At these low atomic numbers, there is a considerable relative difference in mass among the isotopes of a given element. Because of that difference, the isotopes of an element exhibitsimilar, but not identical, behaviour during reactions or phase changes. For any given chemical process, the ratio of the stable isotopes (for example, carbon-13 to carbon-12) in the reactant will differ from that of the resulting product in ways that we can generally predict. Consider the growth of plants. During photosynthesis, the new organic matter is enriched in carbon-12, relative to the atmospheric carbon dioxide from which it was derived. We can measure these differences in isotope ratios in the laboratory using a mass spectrometer. Most importantly, these mass-dependent changes can be related quantitatively to properties of the reaction, such as its temperature and/or reaction rate. They can also be related to the origin and evolution of the reactants and products, according to an increasingly well-understood set of chemical and physical principles.

In the stable isotope business, you are what you eat.

Stable isotope science largely began in the late 1940s, and was most prominently developed in the geological sciences over the next forty years. Many scientists with strong roots in Canada played a founding role in the development of the field, including Sam Epstein, Bob Clayton, and, of course, the late Harry Thode, who was president of McMaster University. Since then, we've had exciting developments. A new approach to instrumentation (continuous-flow mass spectrometry), developed over the last few years, is now causing a second, explosive evolution in the applications of stable isotope methods.

The most dramatic developments are occurring in biology, ecology, environmental science, and medicine. Specific compounds from complex biological materials can now be isolated and analysed from small samples with relative ease. This means that stable isotopes of oxygen, carbon, hydrogen, and nitrogen are now being used for everything from the detection of the origin and adulteration of alcohols (is your favourite libation really what you think it is?), to tracking the migration of monarch butterflies. As well, scientists, motivated by the spectre of mad cow disease, are testing combined stable carbon, nitrogen, hydrogen, and oxygen isotope ratio measurements of hoof and horn as a potential means for identifying the origin of beef cattle. Will it work? It just might. Who could have imagined that such an application of research—originally aimed at relating changes in the composition of the Earth's oceans to variations in the Earth's climate—could one day be used in detective work involving cows?

To get a better understanding of how stable isotopes work, let us consider the water molecule. The water molecule illustrates how stable isotopes can be used to trace important interactions in nature and to indicate how the isotopic signals carried by the water molecule can be incorporated into other materials, which then preserve a record of ancient climates. This will lead you into a maze at the centre of which are Elizabeth Webb (a recently graduated Ph.D. student at the University of Western Ontario) and me. It is a story of an incredible journey of thirty-five-thousand kilometres back and forth across the Great Plains of North America to collect native prairie grasses. This was a great trip, with wonderful scenery and interesting discussions at border crossings concerning a white panel truck stuffed full of grass plants. It was hard work, so why did we bother?

Water is made up of the stable isotopes hydrogen-1, hydrogen-2, oxygen-16, oxygen-17, and oxygen-18. The Earth's main reservoir of

Stable isotopes of oxygen, carbon, hydrogen, and nitrogen are now being used for everything from the detection of the origin and adulteration of alcohols (Is your favourite libation really what you think it is?), to tracking the migration of monarch butterflies.

water is the oceans. The rain and snow that provide our fresh water, whether in lakes, rivers, or groundwater, are ultimately created by evaporation from the oceans, followed by the transportation of this moisture onto the continents. When water evaporates, molecules containing the isotopes with lower masses (for example, oxygen-16) are preferentially transferred to the water vapour in a fashion that can be related quantitatively to temperature and to the rate of evaporation. When this moisture moves across the continents, it is gradually withdrawn from the atmosphere as precipitation. The resulting rain and snow obtains isotopic compositions that we understand and can predict reasonably well.

The end result in North America is a truly amazing, systematic variation in the oxygen and hydrogen isotopic compositions of fresh water. This is strongly correlated with latitude, altitude, and climatic conditions such as temperature and relative humidity. For example, the oxygen and hydrogen isotopic compositions of fresh water in Edmonton are very different from those in London, Ontario. If I were to analyse your body water, it would demonstrate that you are what you drink. An Edmontonian's body water will have a different stable isotopic composition than will that of a Londoner. If you moved from Edmonton to London, your body water would gradually shift, over some weeks, to the isotopic composition of the water in the new location. The oxygen contained in your tooth enamel, however, would not change nearly so rapidly. For the most part, it would preserve an oxygen isotopic composition characteristic of the water at the location where you had lived when those teeth developed. This isotopic signature has been helpful in determining the origins of sacrificial victims in ancient tombs (were they homegrown heroes or captives from the highlands?), and it also has applications in modern forensic work.

But I digress. Back to our trip across the plains and our white truck full of grasses. As vegetation grows, it uses the rain that falls at a particular location. That rain has its own, unique isotopic composition. Before the intensive agriculture of the North American Great Plains, a principal type of vegetation was native prairie grasses. It happens that these grasses also take up dissolved silica from the water in the soils. This silica becomes concentrated in the plant water, and ultimately crystallizes as opal within some cells of the grass plant. These biogenic opals, otherwise known as phytoliths, obtain an oxygen isotopic composition that is determined by the soil water delivered to the plant through its roots, by the growing season temperature, and, in the leaves and flowers, by relative humidity. In other words, each grass plant deposits its own built-in isotopic "silica memory chip" that integrates the temperature, relative humidity, and water composition during plant growth. On our cross-continent trek, Dr. Webb and I were able to establish that principle for plant opals from a wide variety of climatic regimes. Of course, the local weather office can also give you this information more accurately, and for a lot less money and work.

The real potential application for this research arises when we consider what happens when the grass plants die. The organic matter will gradually decay in the soils, but the plant opal will remain. Layer by layer, the phytoliths accumulate in the soil and may persist for up to 100,000 years. Their oxygen isotopic compositions may be a proxy for past climate, provided that it is possible to accurately date the ancient soil horizons that contain the phytoliths. That should be possible. When plant opals are forming, they trap organic carbon. It may be feasible to use the radioactive carbon-14 content of a plant opal assemblage to determine its age. The stable ratio of carbon-13 to carbon-12 in this occluded carbon will also contain information about climate history. The exact amount by which carbon-12 is preferred over carbon-13 during photosynthesis depends on the reaction pathway. Cooler, moister conditions are commonly associated with quite different carbon-13 to carbon-12 ratios than warmer and/or more arid conditions.

Will this new approach be successful? Will there be sufficient resolution using the opal phytoliths to obtain a useful record of ancient climate in particularly sensitive areas such as the Great Plains of North America? Will we be able to deduce the rates of ancient climate change well enough that they can be compared meaningfully with current trends? We don't know yet. Our job is to continue to strive, to ask the right questions, to make the best experiments, and to follow them wherever they lead us.

Research is that perpetual quest. Mine started with a search for gold, which I've never found. But then again, perhaps I have.

Further Reading

"Stable Isotopic Constraints on Sandstone Diagenesis in the Western Canada Sedimentary Basin," Fred J. Longstaffe, in *Quantitative Diagenesis: Recent Developments and Applications to Reservoir Geology*, edited by A. Parker and B.W. Sellwood (Boston: Kluwer Academic Publishers, 1994), pp. 223–274.

"An Introduction to Stable Oxygen and Hydrogen Isotopes and Their Use as Fluid Tracers in Sedimentary Systems," Fred J. Longstaffe, in *Fluids and Basin Evolution*, edited by T.K. Kyser (Mineralogical Association of Canada Short Course Series, Vol. 28, 2000), pp. 115–162.

Stable Isotope Geochemistry, 4th ed., J. Hoefs (Berlin: Springer-Verlag Berlin Heidelberg, 1997).

"The Oxygen Isotopic Composition of Silica Phytoliths and Plant Water in Grasses: Implications for the Study of Paleoclimate," E.A. Webb and F.J. Longstaffe, *Geochimica and Cosmochimica Acta*, Vol. 64 (2000), pp. 676–780.

Dr. Catherine Kallin was born in Grand Forks, British Columbia, and grew up in the Vancouver area. Her childhood background of welfare, foster homes, and day-to-day survival was an unlikely preparation for a life in academe. Eventually, it was her interest in abstract mathematics that drew her to the science program at the Vancouver community college, where, in a "Physics for Poets" course, her teacher passed on to her his love of physics. It changed the trajectory of her life.

In her second year at college, Dr. Kallin transferred to the mathematics and physics program at the University of British Columbia, from which she graduated in 1979. She spent a year at IBM in New York, working on the Josephson Computer Project, a project aimed at developing a new computer based on superconducting devices. She then began graduate studies in physics at Harvard. In 1984, she obtained her Ph.D. in physics and spent the next two years as a post-doctoral fellow at the Institute for Theoretical Physics in Santa Barbara, California, where she worked on the theory of the fractional quantum Hall effect.

In 1985, Dr. Kallin and her husband, a theoretical physicist on the faculty at UBC, decided to end their long-distance commuting relationship and look for jobs that would allow them to live together in the same place with their newborn daughter, Ann. As a result, they both joined the faculty at McMaster University in 1986.

Her research accomplishments have been recognized with Alfred P. Sloan, John Simon Guggenheim, E.W.R. Steacie, and American Physical Society fellowships. She has served on a number of international scientific boards, and has recently returned to McMaster University from Santa Barbara, where she co-organized a five-month-long workshop on high-temperature superconductivity.

High-Temperature Superconductivity

Catherine Kallin

I was first drawn to science through mathematics. Inspired by the beauty, elegance, and power of pure mathematics, it was my desire to apply this power to understanding, at a deep, fundamental level, how the world works that attracted me to theoretical physics. Many students who enter physics through this route end up in cosmology or particle physics, but I chose condensed matter physics (the physics of metals, semiconductors, insulators, plastics, and magnets). Although it may not sound as romantic as other specializations, it is more closely attuned to the world we inhabit.

My journey began in the early eighties, when I spent four years studying condensed matter physics at Harvard University. I continued for two post-doctoral years at the Institute for Theoretical Physics in Santa Barbara before I ended up as an assistant professor at McMaster University in Hamilton in 1986. At the time, I had never worked on superconductors. Instead, I had been studying the

two-dimensional electron gas in semiconductor heterostructure devices.

At just about that time, in early 1987, high-temperature superconductors burst onto the physics scene—in spite of the fact that the actual discovery was published in an obscure German journal in the spring of the previous year. The 1987 announcement was greeted with scepticism. To a large extent, no one was willing to believe yet another observation of superconductivity at an unexpectedly high temperature in a complicated, poorly characterized system. The announcement only became an accepted "discovery" when other labs started to reproduce the results. They wanted to determine the nature of the material that gave the superconductivity, and to find even higher-temperature superconducting compounds. News of these developments travelled fast. When we heard of them, it was enough to launch my colleagues and me in a new direction. This area of research has captivated us for the past fourteen years.

Superconductivity itself is a rather old phenomenon, but a spectacular one nevertheless. At low temperature, the electron liquid that makes metals act as good electrical conductors condenses into a quantum superfluid state of molecule-like "Cooper pairs" of electrons. Supercurrents of these condensed pairs can flow forever without slowing down. The ability to carry electric current without any resistance is what makes superconductors useful for the transmission of electric power. Persistent currents make it possible to have superconducting magnets that consume no power (except for the power required to keep them cold). Such magnets are used for magnetic resonance imaging (MRI).

An even more spectacular consequence of superconductivity is the Meissner effect. When a metal sitting in a magnetic field is cooled into a superconducting state, the magnetic field is expelled. This makes a magnet, sitting on top of a piece of metal that becomes superconducting, jump up and levitate. We already see this principle at work in high-speed Maglev trains, which rely on the "magnetic levitation" created by the Meissner effect. By using powerful electromagnets, these trains float, or "levitate," over a guideway, replacing the old steel wheels and tracks.

Superconductivity is an intrinsically low-temperature phenomenon. Its discovery by Nobel Prize-winning physicist Heike Kamerlingh Onnes in 1911 was a consequence of his earlier work, investigating the properties of matter at low temperatures. This led to the production of liquid helium, and to its subsequent use as a refrigerant. The first superconductor, solid mercury, becomes superconducting at a so-called "critical temperature" (T_c) of 4.1 degrees above absolute zero (written as T_c = 4.1K). Of course, what is high and low in physics is all relative. Astrophysicists study superconductivity in the cores of neutron stars where the temperatures can be billions of degrees. The difference lies in the strength of the attractive forces that bind the Cooper pairs. In solid mercury or lead, the forces between electrons that give rise to superconductivity arise from small, transient distortions of the crystal lattice as electrons whiz by each other. This so-called "phonon mechanism" was first explained by J. Bardeen, L.N. Cooper, and J.R. Schrieffer in their BCS Theory of conventional superconductivity. In neutron stars, the forces responsible for superconductivity are strong nuclear forces.

When a metal sitting in a magnetic field is cooled into a superconducting state, the magnetic field is expelled. This makes a magnet, sitting on top of a piece of metal that becomes superconducting, jump up and levitate. We already see this principle at work in high-speed Maglev trains.

All of the terrestrial superconductors that were known before 1986 were believed to result from the phonon mechanism. The fact that their critical temperatures were all rather low was attributed to the fact that this mechanism is intrinsically quite weak. The observation of superconductivity above thirty kelvin suggested that another, stronger mechanism was at work. The discovery the following year by Paul Chu and his co-workers of superconductivity above ninety kelvin dramatically confirmed this suggestion. Since then, critical temperature (T_c) has been pushed as high as 160 kelvin (–113°C). It is both surprising and puzzling that the class of compounds first studied by Nobel Prize-winning physicists J. Georg Bednorz and K. Alexander Muller are such good superconductors. These materials are brittle, black ceramics with a very low density of conduction electrons. One has to wonder how such bad metals become the world's best superconductors when they are cooled down.

I am a theoretical physicist, so I am not involved in the design of wire, electromagnets, or trains. My interest is in the quantum many-body problem: how 100,000,000,000,000,000,000,000 electrons in a cubic centimetre of metal move through a background of metallic ions, while at the same time interacting with each other through the Coulomb Force (like charges repel, unlike charges attract). This problem has many solutions, depending on the kind and arrangement of the ions, and on the density of the moving electrons. The solutions can involve ferromagnets like iron, normal metals like copper, or superconductors like mercury or lead. One can even have solutions where the electron splinters into pieces, each carrying a fraction of the electric charge of the electron. This happens in the so-called "fractional quantum Hall effect," which is observed in certain semiconductor devices subjected to a large magnetic field. This is a problem I worked on during my postdoctoral years. Similar exotic solutions involving splintered electrons have been proposed for high temperature superconductors. However, unlike the quantum Hall effect, high temperature superconductivity remains an unsolved problem. We do not yet know exactly what kind of theoretical solution describes the electron motion in these materials.

The reason why the theory of high-T_c superconductivity is so difficult is that the forces that are required to give high critical temperatures are very strong. The strong force between electrons is the Coulomb Force, which is repulsive, whereas, as mentioned earlier, superconductivity results from attraction. Cooper pairs in conventional, low-T_c superconductors are weakly bound by lattice distortions. The wave function for this pair is very large, and the Coulomb Force is negligible because the two electrons are mainly far apart. In high-T_c materials, the Cooper pairs are tightly bound and very small. As a result, the two electrons need to work harder to avoid each other. The way they do this is described as a "d-wave" wave function, which is similar to the d-orbital in an atom. This

Although there may still be room for a solitary Einstein to discover the laws underpinning the universe purely through the power of thought, more and more of what we know is being discovered by people working together.

orbital has the property that the two electrons are never in the same place. The possibility of d-wave superconductivity was an early prediction of "strong correlation" theories, in spite of the fact that this kind of superconductivity had never before been observed. The prediction of d-wave Cooper pairing was not a complete theory, but was rather a possible outcome of several classes of theories. Nevertheless, it was important to know whether high-T_c superconductors behaved in this way.

The first compelling evidence that superconductivity in high-T_c materials is d-wave was obtained by three of my University of British Columbia colleagues: Walter Hardy, Doug Bonn, and Ruixing Liang. In 1993, we were working together, trying to understand the microwave conductivity of high-T_c superconductors. That was when it became obvious that we needed to know the temperature dependence of the density of superconducting pairs. In conventional, low-T_c superconductors, the superfluid density hardly changes at the lowest temperatures; then, at temperatures above about half of T_c, the density drops rapidly, falling to zero at T_c. We had already realized that the superfluid density of high-T_c materials was too temperature-dependent at low temperatures. Hardy designed an exquisitely sensitive apparatus to measure this quantity, and the UBC group then discovered, to their amazement, that the superfluid density fell nearly linearly as the temperature increased from absolute zero ($T = 0$) to T_c. This was a remarkable result. It was totally different from the behaviour in conventional superconductors. It also just happened to be exactly the behaviour predicted for a d-wave superconductor. The precision of the UBC measurements was so great, and the comparison between high- and low-T_c materials that they presented was so striking, that physicists around the world were soon convinced that high-T_c superconductivity was d-wave.

I have benefited tremendously over the years from collaborating with these extremely talented and successful experimentalists. I have tried to give something back by occasionally suggesting things that they could measure, by helping with the analysis of their data, and by relating the results to current theories, both my own and those of others. In physics, as in many occupations, effective collaboration is a proven way of getting results. Although there may still be room for a solitary Einstein to discover the laws underpinning the universe purely through the power of thought, more and more of what we know is being discovered by people working together. By combining their different talents and expertise, and stimulating each other to do the best they can, scientists in general, and physicists in particular, have become more social in nature. This is one aspect that I find particularly pleasing.

In the early sixties, when Bob Schrieffer went to the University of Illinois in Urbana-Champagne to work with John Bardeen to solve the problem of superconductivity, that problem had been picked over by generations of theoretical physicists. The level of understanding was very highly developed, and a great deal of useful phenomenology had been derived. People like Frolich and Peierls had even looked at the possibility that phonons were involved. They just couldn't figure out how to make it work. The difficulty was that it took a rather large mental leap to go from standard electron many-body physics to the radically new and different quantum wave function invented by Bardeen, Cooper, and Schrieffer to explain conventional superconductivity.

Now, at the beginning of the twenty-first century, the understanding of high-temperature superconductivity is in a similar state. Many elegant theories have been put forward. Some of them involve mathematical models too difficult to fully solve, and one can only speculate about whether they contain the right physics. Others that can be worked out clearly do not have all the right stuff. The field is waiting for a breakthrough that could come about from inspired thought, or be prompted by new experimental observations. For this reason, it is a very exciting time in the field of superconductivity. It is also a very unsettled one. It is at times

like these that great discoveries are made that have the potential to alter the path of science in ways too difficult for anyone to foresee.

Further Reading

"High Temperature Superconductivity," edited by C. Kallin and J. Berlinsky, *Physics in Canada*, Vol. 56, No. 5 (2000), pp. 217–74.

"Correlated Electron Systems," *Science*, Vol. 288, No. 5465 (2000), 461–82. A special issue of *Science* with an introduction and review of articles on high-temperature superconductivity.

Information on power applications of high-temperature superconductors is available online at: http://www.ornl.gov/HTSC/htsc.html.

Dr. Marc-André Sirard started his career as a veterinarian, a decision influenced by his love of animals and his view that the profession was a noble one. During his years as a farm veterinarian, he became interested in reproduction and decided to pursue a master's degree in a subject that was very exciting at the time: in vitro fertilization. This interest turned into a Ph.D. and approval of the first clinical method to produce test-tube calves. He then decided to devote his career to the study of the oocyte, a "wonderful and immortal cell."

Dr. Sirard obtained his degree in veterinary medicine from the University of Montreal in 1981, and his Ph.D. from Université Laval in 1986. Today, he is a full professor in the Department of Animal Sciences at Université Laval. Since 1988, his laboratory has published more than 125 scientific papers, presented more than three hundred communications, and has been invited to give more than twenty-five lectures at meetings and seminars around the world. He greatly enjoys working with graduate students and has supervised the work of twenty-five master's students and eighteen Ph.D. candidates.

In 1990, Dr. Sirard obtained an Industrial Research Chair in collaboration with Semex Canada, which culminated in the 2000 Synergie Award. He was also one of the first candidates to receive a Canada Research Chair. He is the founder and director of the Research Centre in Reproductive Biology, and is co-founder of TGN Biotech, a biotechnology company based in Ste-Foy, Quebec.

The Ovum's Secret

Marc-André Sirard

No matter what the age of the parents, the baby is always brand new.

Admittedly, that is not an original observation. In fact, for every human who has taken on the biological task of creating life, the outcome is predictable and lacking in mystery. Whether the parents are teenagers, young adults, middle-aged, or even older, the babies are always born young and new. The real mystery, however, begins to develop when we ask ourselves one fundamental question. Why?

Why is it that out of something old, something completely new can be born? What is it about the few germ cells deep inside our bodies that allows them to remain young and vibrant? How is it possible for these few cells—known better as ova and spermatozoa—to remain un-aged in our gonads when all other cells in our body can survive for only a limited number of years? What is the secret that makes these cells potentially immortal? If we

can answer these questions, then we swing the door wide open for biological and medical developments that have the potential to change, shape, and save lives.

In fact, we are discovering that the ovum is not only capable of staying young inside the body until ovulation, it can also erase the effect of time on other cells from adult tissues. The birth of Dolly, the now-famous cloned sheep, has shown us clearly that the ovum is capable of reprogramming adult cells to produce a viable embryo, and that it is the only cell with the right formula for rewinding the biological clock. Although cloning has been repeated in many animal species since then, the molecular mechanism involved in the reprogramming remains a mystery.

A major reason to explain the absence of information about the molecular events happening in the ovum is the fact that mature ova are produced only once per sexual or menstrual cycle (twenty-eight days in humans, compared to four days in a mouse). Furthermore, the ova are about half the size of the period at the end of this sentence. As a result, there is not enough material for chemical or protein analysis. Surprisingly, there is more knowledge of the genes involved in cancer induction than of those associated with the reprogramming ability of the ovum.

Despite all the mystery, there is one species where much progress has been made in our understanding of the ovum and of early embryonic life: the cow. Mainly for genetic purposes, embryo transfer and in vitro fertilization techniques have been developed and used in cows for more than twenty years—and with very good success.

I first became interested in cow reproduction as a veterinarian in the early 1980s, when I was working in a large-animal practice in Lacolle, a small town south of Montreal, near the United States border. After a few years, I decided to specialize in reproduction and to learn embryo transfer. I joined the CHUL Research Center, affiliated with Université Laval in Quebec City. My project involved using bovine in vitro fertilization as a model for human test-tube babies. After two years of working very hard to get ova from cows through laparoscopy (using an optic tube through the flank of the

standing animal on local anaesthetic), and sperm from bulls through manual collection over a collaborating female, I was about to quit. Not one of the ova would start to divide. Then we had our first success. It came when we decided to use heparin to prevent coagulation in the long tube we inserted in a cow to get the eggs. Heparin was later recognized as a functional agent to make the sperm more fertile in vitro. It was a stroke of luck.

Our second stroke of luck came from the use of fallopian tubes from rabbits to culture our bovine embryos. Since the mammalian embryos stay smaller than a dot for seven days, we were able to put more than twenty of them into a single rabbit for five days without any problem. This clinical method, in combination with the laparoscopic recovery, allowed us to obtain the first nine test-tube calves in 1985. After six days of hormonal treatments, the number of eggs obtained from an ovarian-stimulated cow is around a dozen. For fertility treatment, that is more than enough, but to understand what makes these ova capable of reprogramming an adult cell, it is far from the necessary number.

In fact, we are discovering that the ovum is not only capable of staying young inside the body until ovulation, it can also erase the effect of time on other cells from adult tissues.

We then tried to use immature eggs from slaughtered animals to produce embryos in vitro for research, as well as for clinical purposes. This way, hundreds of ova could be obtained with a few hours of work. In two years, during my postdoctoral stay at the University of Wisconsin-Madison, we were able to obtain one bovine embryo for every three ova put in culture. This was a big improvement in the ratio. The next ten years were devoted to the understanding of ovum quality, since only one out of three was good, even if they all looked alike under the microscope. The bovine ovary selects only one follicle per cycle for ovulation. The selection phenomenon, called “dominance,” is based on the ability of the largest follicle to inhibit the growth of the smaller ones. As a result, not all the fully grown ova present at a given time will share the same ability, or competence, to induce embryonic development following fertilization. It seems that the ovary has the capacity to select the best ovum, and before making it competent to form an embryo, it waits for the expulsion or ovulation mechanism to be in place.

During the past several years, research done in our laboratory at Université Laval's Faculty of Agriculture on the ovarian factors influencing the competence of the ovum has led to the development of new bovine infertility treatments. As a result, the rate of success for these treatments now surpasses inherent fertility. We have also introduced a new concept of ovum competence, and are now connecting it to the genes involved in this miraculous ability. But this research, and the resulting knowledge, has applications that go far beyond the bovine world. We are discovering that it can, without a doubt, be used to solve some problems of human infertility. Not surprisingly, gynaecologists from around the world are now interested in our new concept of egg competence.

In addition, recent technological advances in molecular biology and genomics have provided very powerful tools that can be applied to reveal the secrets of the ovum. For instance, it is now possible to identify an individual from a drop of blood or a single hair. What's more, a molecular photocopy machine—derived from a bacterial enzyme with a unique resistance to heat—can copy a gene millions of times, amplifying it sufficiently for analysis or identification. How can we use these new molecular tools to understand how the ovum can erase time? It is now possible to copy all the genes being expressed in the ovum, and then compare them to the genes expressed by other cells. Since similar genes have the ability to bind to each other (if a tenfold excess of genes from skin cells are used to mix with ovum genes), those that are similar will join, leaving the genes that are unique to the ovum. We have adapted this molecular subtraction method to very small amounts of tissue and used it to identify genes specific to the ovum.

Another useful technique is the DNA array, in which thousands of copies of different genes are glued on a glass slide, with each gene in a unique and precise spot. This way, we are able to mix all

If we could find the ova's secret weapon to destroy free radicals (which are gradually toxic to all cells, but especially brain cells), we might find a way to prevent Alzheimer's disease.

of the genes expressed in the skin, and apply the mixture on a slide where thousands of ovum genes have been glued. In a one-step experiment, we can see all the genes common between the two cell types and, more importantly, the ones that might be specific to the ovum. It should be noted that many genes, like the cyto-skeleton or the respiratory enzymes, are common in most cells.

Keeping in mind the similarities among the genomes of mammals, we can say with an amount of certainty that most of the genes present in the cow ovum are probably also present in the human ovum. With the cow, we now have enough ova to explore the genes involved in tissue reprogramming, or even in the ovum's capacity to resist the effects of time. With the recent progress in the Human Genome Project, thousands of human genes are now being discovered. For many of them, we have no clue as to their function. By combining these two sources of information, we will probably be able to identify the human genes involved in these processes.

The information we are now gathering from the cow ovum will tell us how the reprogramming system works. Understanding this reprogramming will permit significant advances in both agriculture and human health. For example, if it were feasible to take a few skin cells from a Parkinson's disease patient and reprogram them to become new nerve cells, they could then be transferred back to the patient to bring about a cure. If we could find the ova's secret weapon to destroy free radicals (which are gradually toxic to all cells, but especially brain cells), we might find a way to prevent Alzheimer's disease. And if we knew how the ovum controls reprogramming, we might be able to envision new treatments to prevent the programming in cancer cells.

At the same time that these applications or technological breakthroughs hold out the promise of cures for human illnesses, they also raise profound ethical questions. If we can lengthen life, should we also reduce reproduction? Who will pay for these tissue-regenerating treatments, or for the making of new organs from old cells?

We believe that these scientific advances have the potential to bring great benefits, but the ethical issues need to be properly defined and addressed. A few years ago, at the Research Centre in Reproductive Biology, we invited more than twenty-five people involved in ethics, law, sociology, philosophy, religion, and animal welfare to reflect on these issues. Over two years, we spent eighteen days discussing our differences of view toward such new developments as transgenic animals and clones. Though we started with different technical languages, we all learned from each other. By the end, we were able to agree on many issues. We also agreed that our next round of study should focus on the reprogramming of human tissues.

We now see the danger of the slippery slope: Something immoral today may become a banality for the next generation. Several years ago, cloning was first banned for reproduction purposes, but it now appears to be viable and acceptable for certain types of stem-cell research. It is clear that scientists alone cannot decide the value of a given technology for society. On its own, knowledge has very little moral value. It is the use of knowledge for a given result that bears ethical value.

Further Reading

"Mammalian Cloning: Advances and Limitations," D. Solter, *Nature Reviews: Genetics*, Vol. 1, No. 3 (December 2000), pp. 199–207.

"Biomedicine—Putting Stem Cells to Work," D. Solter and J. Gearhart, *Science*, Vol. 283, No. 5407 (March 5, 1999), pp. 1468–70.

"Ethical Issues in Embryonic Stem Cell Research," R. Dresser, *Journal of the American Medical Association*, Vol. 285, No. 11 (March 21, 2001), pp. 1439–40.

"A Legal and Ethical Tightrope—Science, Ethics and Legislation of Stem Cell Research," A. Colman and J.C. Burley, *EMBO Reports*, Vol. 2, No. 1 (January 2001).

As a student, *Dr. John Robinson* spent nine years learning about environmental problems and how to do interdisciplinary research. In the process, he received an undergraduate degree in geography from the University of Toronto in 1975, a master's degree in environmental studies from York University in 1977, and, eventually, a Ph.D. in geography from the University of Toronto in 1981. He then spent another eleven years trying to apply some of this knowledge in the Department of Environment and Resources Studies at the University of Waterloo.

The interdisciplinary culture at Waterloo fostered Dr. Robinson's research and teaching interests in a number of areas, including energy efficiency and renewable energy in Canada, socio-economic modelling, forecasting and futures studies, the use of scientific information in decision-making, sustainable development theory, and environmental philosophy.

In 1992, he moved to the University of British Columbia, where he is currently director of the Sustainable Development Research Institute (SDRI) and a professor in the university's Department of Geography. He teaches environmental studies courses at UBC, but spends most of his time creating research projects on a wide range of sustainable development issues. At SDRI, he currently directs several research programs in the areas of climate change and policy, gaming and modelling sustainable futures in the Georgia Basin, and the development of modelling and scenario analysis tools. His personal research interests include the human dimensions of global change, involving the public in the regional analysis of sustainable futures, and the relationship between science and decision-making.

In addition, Dr. Robinson is a convening lead author of the Intergovernmental Panel on Climate Change, a member of the Scientific Advisory Committee of the Inter-American Institute for Global Change Research, and a member of the Advisory Council of the David Suzuki Foundation. He also sits on the editorial board of the journal *Integrated Assessment*.

Changing the World: The Subversive Appeal of Sustainability Research

John B. Robinson

In the fall of 1970, I sat with a group of grade thirteen classmates in Port Hope, Ontario, listening to Monte Hummel from Pollution Probe give a talk on the environment. He held up a strip of cardboard with five squares along its upper edge, ranging in colour from very pale grey to dark black. If the smoke from a factory smokestack was darker than the middle square, he said, then we should phone the Ontario Air Management Branch and report a violation of its air quality regulations. That talk was my introduction to environmental issues and the beginning of a long voyage of discovery.

Thirty-two years later, the environmental argument has become somewhat more complex than just squares on a strip of cardboard. What began as a concern about mainly local, point-source pollution, wilderness conservation, and resource depletion, has turned into an awareness of the potential breakdown and reconfiguration of global biogeochemical cycles, of fundamental and potentially irreversible transformations

in ecological systems, and of massive impacts on human societies. Clearly, our world has changed. This growing awareness of the huge scale and scope of environmental problems has been accompanied by a recognition that human and natural systems are inextricably intertwined, and that there exist multiple feedbacks between socio-economic and cultural circumstances, such as poverty or affluence, and environmental conditions. Moreover, it is becoming clear that the border between the natural and the human is not a stable boundary. Rather, it is a shifting borderland in which technological change, together with changes in our profoundest understandings of the social construction of culture and nature, are altering the ground on which we stand, both literally and metaphorically.

These fundamental transformations of the environmental argument are symbolized by a shift in language. What we used to describe as environmental problems are increasingly seen as issues of (un)sustainability. A recent example can be seen in the work of the Intergovernmental Panel on Climate Change (IPCC). Previous IPCC assessments have focused their attention on the likelihood of human-induced changes in the physical climate system, on the potential impacts of those changes, and on the availability and costs of adaptation and mitigation options. In the IPCC's third assessment report, the links between climate change and sustainable development are receiving much more attention. It has been suggested, for example, that the choices we make regarding our underlying socio-economic and technological development paths may even be more important for climate change effects, adaptation capacity, and mitigation potential as the choice of climate policies. This amounts to a fairly radical recasting of the argument.

The move from "environment" to "sustainability" is a profound one, and I believe the full implications of such a shift have not yet been fully grasped. In what follows, I will try to suggest some of the more challenging dimensions of this new approach. In so doing, we will consider a kind of research and action agenda, together with some indication of how we are trying to address this agenda in the work that my colleagues and I are conducting

at the Sustainable Development Research Institute (SDRI) at the University of British Columbia. A basic premise is that we need significant progress in both the theoretical and applied dimensions of this field. This implies an interactive process of alternating between the difficult rethinking of the underpinnings of our interactions with the natural world and attempts to engage with the world around us.

A traditional view of the relationship between human and natural systems sees both as physical systems that consume energy and alter the material characteristics of their environment. Key issues in this approach have to do with resource availability, emissions, habitat alteration, biodiversity preservation, and potential physical limits to growth. This is the realm of environmental science and quantitative social science.

A quite different view of this relationship focuses on flows of information between the two systems. In what might be called an "actor-system" view, the key questions that are studied have to do with how different theories and explanations about environmental problems get constructed, and how they play out in the realm of decision-making and policy-making. Different interpretations of social and environmental problems are developed and manipulated by different actors in a complex process of social and political negotiation and interaction. This perspective tends to be that of the more qualitative social sciences and humanities.

The point, of course, is not that one view is superior to the other, but that any attempt to come to grips with how humans live on earth must encompass both kinds of interaction. We clearly need the insights of, say, atmospheric science as to how the physical climate system works if we are to address climate-change problems. Just as importantly, if we are to have any chance of implementing meaningful policy, we need to understand how various actor systems construct and address the climate-change problem, and how it plays out in different institutional contexts. It is this double-barrelled perspective, ranging from rigorous empirical science to postmodern deconstruction of human knowledge systems, that the challenge of sustainability demands.

In conjunction with a few research centres in Europe, at SDRI we are attempting to grapple with this requirement. We are developing research programs and computer modelling tools that embed detailed, quantitative models of natural and human systems within qualitatively rich "user interfaces" that are centred on storytelling metaphors and that require an explicit treatment of political process.

Traditional environmental research is often driven by science. That is, the problem definitions and approaches emerge out of the disciplines (ecology or economics, for example) used to study the problem. Often, the difficulty is that these definitions and problems do not connect very well with real-world decision-making processes or institutions.

An alternative approach is to start from real-world problems as expressed by actual decision-makers, and work back to the scientific understanding needed to address those problems. As an example, many developing countries resist the view that climate change—a classic science-driven problem—is something they should or can respond to. However, framing climate-change questions in terms of sustainable development issues already on the agenda in those countries may engender a more positive response. It also suggests new ways to think about policy options. While it may threaten traditional hegemonies of knowledge and power over the research agenda, it also highlights the need to pay serious attention to the way in which problems actually manifest themselves in the world.

This insight finds expression in our own research in "interface-driven modelling," which works in the opposite direction from much traditional modelling.

In our work, science-based models do not drive, but are instead driven by, the design of the user interface. In turn, the design is based largely on a process of consultation with user groups. In this way, the models speak directly to the concerns of the users.

While support for greater interdisciplinarity is almost universal in academic discourse today, it is arguably much more rhetorical than real. Even when different disciplines are involved, true

interdisciplinary teams that work intensively together are still rare, largely because the academic reward system is still strongly disciplinary.

An alternative approach recognizes the importance of being critical, in principle, of the very idea of disciplinarity. Sometimes we need work that challenges the very basis of disciplinary understanding. Perhaps there can even be learnable expertise in being able to integrate thinking across different fields of study. At SDRI, all our projects involve collaborative, interdisciplinary teams and, in each one, we attempt to have active representation across all three major funding councils. We also actively encourage graduate students to work across disciplinary lines.

Many developing countries resist the view that climate change—a classic science-driven problem—is something they should or can respond to.

Sustainability research inherently points to the future, and to changes in the way things are done. But while it may make sense to predict the future state of non-human systems, human society is characterized by the existence of intentionality and choice. The goal should not be to predict the most likely future for human systems, but to explore the desirability and feasibility of alternative future configurations for the various systems we care about. This leads to a particular approach to modelling and analysing such systems, and it takes us in the opposite direction from the predictive modelling of a type familiar to the various natural and social sciences. It also forces explicit treatment of the issue of normative choice, and makes it more difficult to obscure such choices with a cloak of ostensible value neutrality. Lastly, it challenges conventional notions of objectivity.

It is not meaningful to talk about "business as usual" when describing the evolution of social systems over a period of decades. Instead, there exists a range of alternative futures, each with individual characteristics, and each more or less subject to various forms of choice and influence, and various constraints. This puts the onus for achieving future outcomes back where it belongs: in the political arena.

Such an approach has significant implications for the types of analysis and modelling that can be done. SDRI has pioneered the development of backcasting models that allow the interested public and experts to explore the physical feasibility, impacts, trade-offs, and higher-order consequences of scenarios designed to achieve the goals and preferences of the user.

Traditional environmental research is often very top-down and one-way in its approach to communication. The goal is to communicate the results of the research to various audiences, who presumably need expert input to improve their decision making. In contrast, an interactive approach recognizes multiple sources of useful knowledge, including lay expertise of various kinds. It also recognizes that key judgments—about what matters, what problems are most pressing, how best to study them, and how to interpret key results—have important ethical and value dimensions. The result is an approach where the "user community" is actively involved in aspects of problem definition, the research process itself, and the interpretation of results. This represents a significant challenge to traditional notions of expertise, intellectual authority, and dissemination.

At SDRI, we encourage the user community (general public, professional groups, interest groups, school children) to actively participate in the design of models and the generation of research results by using our modelling tools to create scenarios that reflect their preferences, values, and beliefs. The result is a kind of fusion of expert understanding with lay values and beliefs.

It is no accident that the widely diverse fields of non-equilibrium thermodynamics and post-structuralist literary theory both agree that the locus of excitement and creativity is not at the centre, but at the edge of the systems we are studying. Whether we want to maximize the chances of making discontinuous jumps to new states of system organization, or to de-centre the text and validate marginal voices, we are expressing a

It is not meaningful to talk about "business as usual" when describing the evolution of social systems over a period of decades. Instead, there exists a range of alternative futures, each with individual characteristics, and each more or less subject to various forms of choice and influence, and various constraints. This puts the onus for achieving future outcomes back where it belongs: in the political arena.

contemporary view of the limitation of an equilibrium-based approach to understanding. In this sense, the work being done at SDRI simply reflects a more general shift in thinking that is increasingly permeating the intellectual culture of our time. One of the key characteristics of this new approach is a form of epistemological reflexivity. It recognizes the degree to which we are influenced by various frameworks and ruling ideas that condition the way we look at the world, and the kinds of problem definitions and solutions that we can come up with.

From this point of view, the trajectory of human concern about environmental issues in this century is part of a larger set of critiques of modern industrial and scientific culture. These critiques emerged in very different ways in the writings of the early Romantic poets and social critics of the eighteenth and nineteenth centuries. Modern concerns about sustainability have deep roots in these critiques. At the dawn of the twenty-first century, they express a desire to rethink our role in nature in a fundamental way. This is not just a matter of changing behaviours and technologies, but of thinking through what it means to be human, what it is to be natural, and what kind of joint technological and socio-cultural world (and world view) we are creating.

In other words, the wider view of sustainability is subversive of both our social practices and of our ways of organizing and institutionalizing knowledge. That is one more reason why our own work at SDRI is focused on building computer-based models to challenge the view that there is only one way of representing reality, and why it is focused on taking these models out to the widest possible audience to engage civil society in a collective search for preferred futures. To those of us at SDRI, that is a pretty exciting research agenda.

Further Reading

"Integrating Climate Change and Sustainable Development," J. Robinson and D. Herbert, *International Journal of Global Environmental Issues*, Vol. 1, No. 2 (2001), pp. 130–49.

"Back from the Future," J. Robinson and D. Herbert, *Alternatives*, Vol. 26, No. 2 (Spring 2000), pp. 32–33.

"Tools for Linking Choices with Consequences," D. Biggs, J. Robinson, C. Tague, and M. Walsh, in *Seeking Sustainability in the Lower Fraser Basin—Issues and Choices*, edited by M. Healey (Institute for Resources and the Environment, Westwater Research, Canada, 1999), pp. 237–62.

"Reconciling Ecological, Economic, and Social Imperatives: A New Conceptual Framework," J. Robinson and J. Tinker, in *Surviving Globalism: Social and Environmental Dimensions*, edited by T. Schrecker (New York: St. Martin's Press, 1997).

Life in 2030: Exploring a Sustainable Future in Canada, J. Robinson, D. Biggs, G. Francis, R. Legge, S. Lerner, S. Slocombe, and C. Van Bers (Vancouver: UBC Press, 1996).

Dr. Jim Miller is a professor of history at the University of Saskatchewan, where he has taught since 1970. Although his teaching and research initially focused on English-French relations, he shifted his focus in the early 1980s to study and teach the history of relations between native peoples and newcomers to Canada.

Dr. Miller's research has produced more than sixty articles and six books, including a general history of native–newcomer relations entitled *Skyscrapers Hide the Heavens: A History of Indian–White Relations in Canada* and a comprehensive history of residential schooling for native children in Canada, *Shingwauk's Vision: A History of Native Residential Schools.*

His scholarly work and other contributions have been recognized in a number of ways. *Skyscrapers Hide the Heavens* and *Shingwauk's Vision* were both awarded prizes. Dr. Miller was elected president of the Canadian Historical Association in 1996–97, and was recognized by the University of Saskatchewan as its Distinguished Researcher at the spring 1997 convocation. He has served as a member of the Social Sciences and Humanities Research Council since the spring of 1998, and in the autumn of 1998 was inducted as a Fellow of the Royal Society of Canada. In April 2001, he became a Canada Research Chair holder in Native–Newcomer Relations at the University of Saskatchewan.

Born and raised in Cornwall, Ontario, he was educated at the University of Toronto, where he received his Ph.D. in 1972. He lives with his wife, Mary, in Saskatoon.

Native–Newcomer Historical Inquiry

J.R. (Jim) Miller

Until the early 1970s, very few academics—whether they were anthropologists, historians, or legal scholars—showed much interest in the history of relations between the indigenous peoples of Canada and European newcomers.

For both sociological and methodological reasons, this neglect of Aboriginal peoples and the history of their interactions with newcomers to their lands began to change in the 1970s. Political and social influences in the 1980s and 1990s accelerated and accentuated this newfound academic interest. Today, we are experiencing a completely changed scholarly landscape. The study of native–newcomer history is flourishing and has proved influential within the academic world and beyond.

Before I continue, I would like to take this opportunity to point out one thing. To some extent, my own situation illustrates the awakening interest in the native–newcomer field since 1970. The son of Scottish immigrants, I was the first of my family to attend university when I went off to the University of Toronto in 1961. Although I came from Cornwall, Ontario, a small city located very close to Akwesasne, one of Canada's best-known Indian reserves, I knew neither native people nor their history. While studying history in those early years, my ignorance did not matter academically; Aboriginal people were almost totally absent from the university history curriculum. The abysmal job that historians did in working Aboriginal people into their subject was considered, by some analysts, to be an indication that they were often not considered to be deserving of serious attention.

It was only after I accepted a position teaching Canadian history at the University of Saskatchewan that an interest in the history of native–newcomer relations—from a teaching and research perspective—began to develop. By the early 1980s, no doubt influenced by my vantage point in Saskatoon, and by the rising evidence of Aboriginal political and legal assertiveness, I was ready to take the plunge into a full-time concentration on native–newcomer history. Among my first works were a general historical survey of native–newcomer history in Canada, a detailed study of residential schools for native children, a biography of the great Cree Chief Mistahimusqua (Big Bear), and with two other researchers of similar background, a history of treaty-making in Saskatchewan. Like many other scholars interested in this field, I also became involved in applied research closely related to my scholarly writing, which has made working as a historian an engrossing and intellectually stimulating experience for most of the past two decades.

In the past, what literature existed on native peoples in the nineteenth century was produced by Christian missionaries who worked among First Nations, or by self-trained amateurs who developed a curiosity about the indigenous population of our still-young country. Scholars, or self-proclaimed scholars, tended to dismiss the contribution of Aboriginal peoples to Canada. When William Kingsford, the so-called "Dean" of nineteenth-century English-Canadian historians, considered including Aboriginal peoples in his

When William Kingsford, the so-called "Dean" of nineteenth-century English-Canadian historians, considered including Aboriginal peoples in his ten-volume *History of Canada*, he decided not to interrupt his narrative. Kingsford concluded that the study of the "Indian races" was "a special field of enquiry, and is totally independent of the *History of Canada*, except so far as it bears upon the relations of the European and Indian races."

ten-volume *History of Canada*, he decided not to interrupt his narrative. Kingsford concluded that the study of the "Indian races" was "a special field of enquiry, and is totally independent of the *History of Canada*, except so far as it bears upon the relations of the European and Indian races."

Law was no better than history. As the Supreme Court of Canada noted in 1990, lawyers virtually ignored Aboriginal peoples and their land rights until 1970. Although anthropologists pursued a scholarly interest in native peoples before the 1970s, their inquiries were motivated by a desire to catalogue aspects of native life before (the way they saw it) a vanishing race disappeared. Consequently, their approach, known as "ethnography," tended to freeze and isolate the societies it examined and ignored their interactions with others.

During the 1970s, scholarly interest in native–newcomer relations began to develop perceptibly because of changes to both academic personnel and research methodologies. The young historical researchers who were moving into the universities were often different from the Caucasian, middle-class males who had previously dominated the academic world. The young academics were often from working-class backgrounds, occasionally were non-European in ancestry, and some were even female. Coming from different backgrounds, they had different interests and different scholarly preoccupations. Their novel concerns are often grouped under the label "new social history," which is an abbreviated way of saying they investigated the poor, the marginalized, and those whom history has ignored. The groups that now attracted researchers' attention were women, the working class, minority ethnic and racial groups, and, of course, the indigenous societies of North America. For their doctoral dissertations, two young historians, Jennifer Brown and Sylvia Van Kirk, took a fresh look at a venerable historical subject—the fur trade. They revolutionized scholarship by approaching the documents with a new question: What role did native women have in the fur trade? When they asked this new question of old documents, they got dramatically different answers. These answers enriched students' understanding of both the fur trade, and of the ways in which natives and newcomers interacted in that commerce.

As Brown and Van Kirk showed, innovative research questions from new researchers often require novel methodology. Sometimes uncovering the role and contribution of the historically inarticulate requires the use of different sources and methods. Consequently, these younger historians began to look at unconventional materials, such as quantitative data in censuses and police records. They also considered oral history accounts, which were the way Aboriginal peoples traditionally kept their histories.

At the same time, researchers into the history of native peoples and native–newcomer relations began to employ a technique known as "ethnohistory," an approach to historical research that used the knowledge that anthropologists had recovered about Aboriginal societies to probe documents produced by Europeans, in search of native experience. For example, the annual reports that Jesuit missionaries sent back to supporters in seventeenth-century France were re-examined to filter out the ethnocentric voice-overs of the frequently prejudiced evangelists. The intent was to allow the native voice to come through more clearly. The ethnohistorical method, and the increasing use of interviews and oral accounts, meant that researchers who had an interest in native peoples had new and effective tools with which to study and report on the groups in which they were interested.

And study they did. Beginning slowly in the 1970s and gathering momentum in the 1980s, academic research on Aboriginal peoples and their interactions with European newcomers grew in both volume and sophistication. At first, the scholarship tended to concentrate on familiar topics, such as missionaries in seventeenth-century New France, or the fur trade. By the 1990s, however, it was expanding to cover almost all regions of Canada and most periods of its history.

This work in history scholarship by latter-day converts was paralleled by the scholarship of their colleagues in anthropology and law. Anthropologists experienced a crisis of professional conscience in the late 1960s as they became aware that the pioneers of their discipline had either diminished Aboriginal societies with their ethnographic surveys, or had been complicit in the development of government policies that worked to the disadvantage of First Nations and Inuit peoples.

By 2000, history, law, anthropology, political science, and sociology could all boast significant expertise in the study of indigenous societies.

Legal scholarship began to correct its long silence about Aboriginal issues in 1970 with the publication of *Native Rights in Canada*, by Peter Cumming and Neil Mickenberg. Through that decade, Aboriginal law, including its historical branch, developed steadily. It was in the 1980s, though, that the field exploded and staid law professors became legal activists. Academic lawyers carried out an enormous amount of research during the long battles over constitutional renewal in 1982 and the ensuing First Ministers' conferences. The protracted efforts to get such Aboriginal rights as self-government recognized in the Meech Lake Accord from 1987 to 1990 and the run-up to the 1992 Charlottetown Accord also produced much research. Usually the purpose of the research was to support the struggle of Aboriginal groups to secure acceptance of their agenda at intergovernmental conferences. It was also to spell out the contents of the Aboriginal and treaty rights that had been "recognized and affirmed" by the 1982 constitutional revision.

By the end of the twentieth century, the result was that scholarly writing on all aspects of Aboriginal life and history, including the history of native–newcomer relations in Canada, had been revolutionized. Before 1970, these fields had been occupied almost exclusively by anthropologists who wrote descriptive ethnography, which tended to freeze Aboriginal people in a certain period, and ignored their interactions with newcomers. But by 2000, history, law, anthropology, political science, and sociology could all boast significant expertise in the study of indigenous societies. Moreover, the approaches that these disciplines took to their subject had also undergone enormous change. Historians and anthropologists employed ethnohistorical methods to re-examine conventional documentation for evidence of native voices. They also relied increasingly on oral-history research to collect stories directly from Aboriginal communities and individuals. In fact, lawyers and political scientists were among the most active and influential scholars, focusing their academic publications on contemporary debates about self-government, land claims, and other issues of significance to Aboriginal communities. Sociologists and some anthropologists were also heavily involved in what is called "action research," which almost invariably meant research designed to support pro-Aboriginal advocacy.

The research revolution in native–newcomer history has had more benefits than just maintaining the interest of researchers in the field. Today, undergraduates in Canadian history classes are offered a much richer and more sophisticated depiction of their country's evolution than was available forty years ago. Social diversity and complexity are major themes in most general accounts of Canada's history. In that regard, the roles of First Nations, Métis, and Inuit receive great prominence. In contrast to what the analysts in Canadian history textbooks noted in 1970—that the "Indian" is not often considered to be deserving of serious attention—newer texts devote considerable space to First Nation societies and their development, as well as to their relations with newcomers.

Today, undergraduates in Canadian history classes are offered a much richer and more sophisticated depiction of their country's evolution than was available forty years ago. Social diversity and complexity are major themes in most general accounts of Canada's history. In that regard, the roles of First Nations, Métis, and Inuit receive great prominence.

The applied research of historians examining native–newcomer relations has also contributed enormously to Canadian public life, particularly since the 1980s. Historians, as well as other scholars whose research deals with Aboriginal peoples, frequently testified in critically important Aboriginal rights cases that were ultimately decided by the Supreme Court of Canada. Less visibly, legions of academic historians and their graduate students have conducted research and testified in land claims cases, as well as in other disputes where the history of Aboriginal peoples' interactions with newcomer populations has been an important consideration. For example, I have testified before the Indian Claims Commission in a Saskatchewan land claim; written reports and been cross-examined in the pre-trial phase of litigation involving a land claim in the Sarnia area; appeared as a witness for a First Nation defendant in a gambling prosecution in Saskatchewan; written expert opinion for a Vancouver law firm representing victims of residential school abuse; given an invited presentation to the Residential Schools Working Group of the federal government's justice and Indian affairs departments; and given dozens of media interviews in which I commented on historical issues involving native–newcomer history.

The revolution that occurred in the writing of native–newcomer history in Canada during the past thirty years is remarkable. In the 1970s, a new generation of historians began to ask new research questions, yielding dramatic new interpretations of many aspects of Canadian history. Their discoveries led to a more complex and richly textured understanding of Canadian history.

Their curiosity-driven research findings facilitated applied research on topics in Aboriginal affairs that have contributed significantly to public, legal, and political debate. There is even evidence that their research findings have begun to influence Canadian popular culture. All in all, it is an exhilarating time to be involved in exploring the history of native–newcomer relations in Canada.

Further Reading

Strangers in Blood: Fur Trade Company Families in Indian Country, J.S.H. Brown (Vancouver: University of British Columbia Press, 1980).

Citizens Plus: Aboriginal Peoples and the Canadian State, Alan C. Cairns (Vancouver: University of British Columbia Press, 2000).

Skyscrapers Hide the Heavens: A History of Indian–White Relations in Canada, 3rd ed., J.R. Miller (Toronto: University of Toronto Press, 2000).

I Have Lived Here Since the World Began: An Illustrated History of Canada's Native Peoples, Arthur J. Ray (Toronto: Key Porter, 1996).

Many Tender Ties: Women in Fur-Trade Society, 1670–1870, Sylvia Van Kirk (Winnipeg: Watson & Dwyer, 1980).

Dr. Robyn Tamblyn is an epidemiologist and associate professor at McGill University's Faculty of Medicine. She has devoted her career to developing and evaluating ways to improve health care delivery. Initially, she did this as a practitioner, educator, and provincial practice consultant. But since 1989, she has accomplished this as a career scientist supported by the Fonds de recherche en santé du Québec, the National Health Research and Development Program, and the Canadian Institutes of Health Research.

Dr. Tamblyn is currently principal investigator at the Quebec Integrated Health Care and Research Network, which is being developed by fifteen scientific teams. The network is headed by some of the province's most outstanding scientists in social, ethical/legal, clinical, health services, and population health research.

Over her career, Dr. Tamblyn has made distinctive and original contributions to science in the areas of prescription drug use and medical education. She has pioneered methods of using data from Canadian prescription claims to evaluate prescription drug use, and has done policy-altering research on the impact of cost-sharing for prescription drugs on prescription drug use and adverse outcomes. The widely publicized results of this research led to targeted modifications in provincial drug policy. Dr. Tamblyn led a multi-centre team of twenty-one investigators through a five-year research program to investigate methods of enhancing optimal drug use and minimizing drug-related illness. In addition, she developed novel computer-based tools to enhance the management of prescription drugs in primary care practice through linking to provincial databases. This initiative will be expanded into fully integrated clinical care delivery systems, led by a number of clinical scientists in primary and specialty care.

The New Millennium Model for Health Care and Research

Robyn Tamblyn

Canada is an international leader in health care delivery, and we're proud to say so. But that enviable status is a fragile one, threatened by the rapid growth in health care spending that is afflicting all industrialized nations. Unless we make fundamental changes in the way we deliver care, the highly valued assets of our Canadian universal health care system—equality of access, and universal and comprehensive coverage—will be lost. Obviously, a challenge of this size demands a strong and coordinated response.

That response is a concept called "e-health" and it has the potential to be as accessible as your computer keyboard. E-health (or *e-santé*) is taking the electronic networking possibilities of the World Wide Web and applying them to patient health care. Along the way, this electronic byproduct of the Internet not only creates a rich resource of patient information that physicians, caregivers, and patients can access twenty-four hours a day, it also aids in the management of clinical and health

record information, and promises to help ring up substantial savings in health care costs.

Developments in e-health are already addressing the problems that currently characterize the state of the health information system—problems that make the system local, idiosyncratic, and fragmented. E-health systems may also hold the potential to spur an exponential growth in our capacity for clinical research.

The idea is already winning fans. Not surprisingly, many governments have seen the great potential for benefits from e-health, and have placed a premium on e-health research—even though the best methods of harnessing the new technologies have yet to be defined. In fact, the situation has created a gap between the concept and the reality. It is this gap that the Quebec Integrated Health Research Network proposes to fill. To develop this research network, our team at McGill University assembled fifteen scientific teams. These are headed by some of the most outstanding scientists in Quebec currently doing research in the social, ethical/legal, clinical, health services, and population health fields.

Over a period of four years, this research network will conduct pilot tests in five areas: primary care, asthma and congestive heart failure, breast cancer, stroke, and transplant surgery. With our tests, we have three main objectives. First, we are working out models of an integrated health care delivery system based on electronic health records and electronic messaging. Second, we are creating prototypes for population-based clinical data registries for scientific research in these five areas. Third, we are developing new health informatics techniques—new ways of using computers to store and organize information—then helping health professionals use these electronic medical records in the most efficient way.

The starting point for this initiative can be traced back to the real-life day-to-day challenges of our health care system. My personal involvement began in the early 1990s with my research on prescription drug use. Using the database of prescription information collected by the province of Quebec, my team members and I found major problems in terms of the frequent issuing of potentially inappropriate prescriptions. The numbers were particularly

high for seniors. A major contributor to this problem was the fragmentation of drug management. Too many different physicians, working independently of each other, were prescribing drugs for the same patient. With no cohesive and organized approach, it was inevitable that problems relating to therapy duplications and drug interactions were inadvertently created. What could be done? We thought about ways to solve the problem and came to the conclusion that one thing was essential: a computerized platform that would allow us to link information and databases with the goal of better managing a patient's prescription information. As a result of that research, we developed new computer-based tools. They would enhance the management of prescription drugs in primary-care practice by linking the computers of physicians and health care providers with provincial databases. Those solutions remain novel to this day. They led directly to the vision of better health care through electronic information management, and today, that is what forms the heart of the Quebec Integrated Health Research Network.

Too many different physicians, working independently of each other, were prescribing drugs for the same patient. With no cohesive and organized approach, it was inevitable that problems relating to therapy duplications and drug interactions were inadvertently created.

Today, our targets are the built-in shortcomings of the current health care system. To begin with, the system has become too unwieldy to meet the main challenge created by an aging population: the management of chronic and complex conditions such as stroke and heart disease. As health care providers and institutions become more specialized, and as newer, high-cost technologies are clustered in regional centres, patient care is increasingly fragmented among multiple caregivers. With each health centre and care provider maintaining separate files, there is often no efficient way for them to communicate, or to share and access essential information. Meanwhile, masses of new information on prevention, diagnosis, and treatment of illness and disease keep flowing in. Few health professionals are able to keep abreast of all the latest developments, even in their own specialties. These include some of the scientific breakthroughs that could benefit their patients.

Let's consider a stroke patient. A stroke is a devastating and disabling event that becomes more common as people age. It is also an illness that requires input and care from a long list of health professionals. After a stroke, a patient is first treated in an acute-care setting. Later on, the patient may require additional care as an out-patient in a clinic, as well as at home. In addition, very complex follow-up management—physiotherapy, occupational therapy, speech therapy—is often needed to reduce disability and optimize recovery. Since the risk of a second stroke is high, follow-up medical management is also critical to prevent recurrence. Several specialists in different centres may be involved, not to mention the primary-care physician who does most of the diagnosis and disease management.

Although our hospitals are well equipped to care for acute stroke patients, unfortunately the story often takes a different turn after patients leave the hospital. Many are left to flounder in the complexities of the health care system. A primary-care physician may not even know where the patient was discharged from, or when. The physician may not be aware of the patient's overall state of health until the patient arrives back in the physician's waiting room, ill and with a pocket full of medications. It is not uncommon for a patient to land back in the emergency room, with another stroke or a different, deteriorating medical condition.

Thanks to an earlier study, however, we didn't have to work too hard to make our case—we already had proof of the value of providing coordinated care *after* a patient leaves a hospital. When a Quebec team devised an early discharge program for stroke patients, they realized that, if the program were to be both safe and effective, patients would need a case manager to coordinate after-care arrangements, organize services, and keep the various physicians in touch. With a manager in place, not only did patients experience a shorter hospital stay, which cost the system substantially less, but fewer patients had to be readmitted to a hospital.

Although our hospitals are well equipped to care for acute stroke patients, unfortunately the story often takes a different turn after patients leave the hospital. Many are left to flounder in the complexities of the health care system. A primary-care physician may not even know where the patient was discharged from, or when.

After that initial success, we realized it would be possible to move on to the next step: creating a *virtual* case manager for each patient, using electronic, computer-based health records. How would it work? Once we have built a complete electronic health record environment—perhaps not for a decade or so—we would simply e-mail a health record summary to the patient, the primary-care physician, the follow-up clinic, and the local community centre for health and social services. Telecommunications messaging systems would be embedded in the health record, and would send reminders to everyone about scheduled treatments. In addition, the caregiver would have access to the latest information about after-stroke management through a health portal.

We're not there yet, but we will be. In the meantime, we're building prototype systems that will be able to automatically send information about the acute-care phase of a patient's treatment to the case manager's personal digital assistant (PDA) and to the primary-care physician by fax. We could also make sure that this information is added to the electronic health care records at the stroke follow-up clinic in advance of a patient's arrival. As we do all this, we are creating electronic consent mechanisms that will meet certain ethical and legal standards. They will also be easily understood by patients, and will allow them to have control over the privacy and confidentiality of their records. These are important issues for which we will develop detailed guidelines. One final, key element: we are working out ways to encourage people to use the system.

Another scenario that demonstrates the need for an electronic networking system arises in prescription drug dispensing—the fastest-growing part of the cost of Canada's health care system. Prescription drugs are powerful agents that can cure or prevent disease, but they can also cause great harm if used inappropriately. Adverse drug effects are now the sixth leading cause of death in the United States. Although Canadian figures are not available, we suspect they are comparable.

To those of us who focus on this issue, it is very clear where the problems lie. The most common prescription drug errors are made in dosing or in transcription. We've all heard of a patient receiving too large a drug dose by accident, or receiving the wrong drug because the pharmacist misread the doctor's handwriting. These things happen regularly, and have caused illness, disability, and even death. Fortunately, such errors are easily prevented with an electronic prescription program. Working with the Canadian Institute of Health Information and a group in Ontario, we are developing an efficient, one-step electronic prescription process that will raise an alert if the drug or dose being dispensed does not conform to normal care. This system should ultimately be available to all physicians, nurses, pharmacists, and patients.

The next step is to set up an electronic record that will inform the primary-care physician of all drugs prescribed for a patient, something we've already pilot-tested. People often don't recall their drugs by name. If they are taking a dozen different medications, it is unlikely they will remember, or tell their doctor about, all of them. Keeping in mind that there are about thirty-three thousand known drug interactions to watch for, as well as some six thousand drugs that should not be prescribed with certain pre-existing health problems (drug contraindications), confusion on the part of the patient and lack of information for the physician can often lead to serious problems. We can also see therapeutic duplication—two drugs from the same class being prescribed by different doctors, or by the same doctor who doesn't have the latest information on those drugs. These problems are especially troublesome for the elderly, who use about forty percent of all prescription drugs. With all of a patient's prescriptions easily accessible in an online electronic record, though, the system can

automatically screen for drug interactions, contraindications, and duplications.

The third issue is poor patient compliance. This is usually caused when a patient misunderstands how a drug should be taken. Taking too much could put the patient in an emergency room, or worse. One way to prevent this is to keep track of drug dispensing, and to alert the physician when a prescription is refilled too often. More commonly, however, a patient skimps on taking a drug they cannot afford—with the result that the disease does not respond well to the prescribed treatment. Surprisingly, some doctors are unaware of drug costs and of their patients' ability to pay. The solution is simple. We feed the cost of a prescription into the electronic system and map it to the amount that is dispensed. If refills are slow, another alert goes out to the physician or caregiver.

In a few years, the electronic health record will be *the* medium to use for conducting health research—which means we won't be spending millions of dollars on labour-intensive data-collection methods.

It is clear that e-health holds the promise of solving many of the problems plaguing our fragmented health care system. But one problem—the current bottleneck in research capacity—overshadows the rest. That is because it affects our ability to develop new knowledge. Two developments have created this bottleneck. One is the escalating demand for research on the efficacy and cost-effectiveness of new tests, treatments, clinical programs, and policies. The other is the high cost of conventional research methods, including labour-intensive recruiting of patients, collecting of data, and also follow-up. How expensive can things get? A major research study can easily cost a million dollars. A major cohort study, like the Canadian health and aging study funded by Health Canada and the National Health Research and Development program, can cost as much as ten million dollars.

We believe such studies will soon be passé, though. In a few years, the electronic health record will be *the* medium to use for conducting health research—which means we won't be spending millions of dollars on labour-intensive data-collection methods. We will soon be able to take research shortcuts by easily retrieving the records of all people who are, for example, asthmatic, following them over time, observing the effectiveness of new drugs and treatments, and recruiting selected persons for special parts of the study. With the integrated health record system in place, our research capacity will increase tenfold, and at one tenth the cost.

Looking ahead, it is clear that we have a big job with a big responsibility. But if we do our work well, the infrastructure we are building will provide a better, more sustainable health care system. With better and faster electronic integration of information, we can bridge some of the gaps where people have been making decisions in the dark. And by embedding information about health research in the electronic health record, we can put the latest scientific breakthroughs into practice now, instead of waiting five to eight years—the current lag time—before new science reaches the patient.

As a scientist, I believe that one enormous benefit of this change is that we will be able to study health and disease in a way that is now sometimes impossible. There has been almost no research done in the management of primary care, or of trauma or emergency care. Although an organized infrastructure has been developed to research the causes of some diseases, like cancer, this is not the case for many chronic diseases like diabetes or chronic heart failure. I feel confident that the Quebec Integrated Health Research Network will place powerful tools right at our fingertips. These tools will help us to better understand the causes and outcomes of diseases, and ultimately the best ways to treat them.

Further Reading

"The Regenstreiff Medical Record System: A Quarter Century Experience," C.J. McDonald, J.M. Overhage, W.M. Tierney, P.R. Dexter et al., *International Journal of Medical Informatics*, Vol. 54, No. 3 (1999), pp. 225–53.

"Developing and Implementing Computerized Protocols for Standardization of Clinical Decisions," A.H. Morris, *Annals of Internal Medicine*, Vol. 132, No. 5 (2000), pp. 373–83.

"Health Care Reform and the New Economy. Does the new digital economy require a different vision for health reform—its principles as well as its possibility?" P. Starr, *Health Affairs*, Vol. 19, No. 6 (2000), pp. 23–32.

After earning a Ph.D. in mechanical engineering, *Dr. Willem Vanderburg* moved to France and continued his study of technology via the social sciences and humanities, under the direction of philosopher and theologian Jacques Ellul. A professor at the University of Bordeaux, Ellul was well known for his theme of the technological tyranny over humanity, and his theories about the "threat to human freedom and Christian faith created by modern technology."

Dr. Vanderburg continues to pursue his interest in these themes and ideas today as the founding director of the Centre for Technology and Social Development in the Faculty of Applied Science and Engineering at the University of Toronto. The centre's goal is to make future engineers more socially and environmentally aware, in order to enable them to create more context-compatible technologies.

His publications include *The Growth of Minds and Cultures: A Unified Theory of the Structure of Human Experience* (1985), *Perspectives on Our Age: Jacques Ellul Speaks on His Life and Work* (1981, 1986, and 1997), "Political Imagination in a Technical Age" (1988), and "Rethinking End-of-Pipe Engineering and Business Ethics" (1997). He also recently completed *The Labyrinth of Technology* (2000), which presents a conceptual framework for examining preventive approaches for technological and economic development. Dr. Vanderburg is also editor-in-chief of the *Bulletin of Technology and Society.*

Preventive Approaches for the Engineering and Management of Technology: Bridging the Gap Between Intellectual Cultures

Willem H. Vanderburg

Although I recognized this only in hindsight, my research into preventive approaches for the engineering, management, and regulation of modern technology began during my days as a doctoral student in engineering.

Discussions about a possible collapse of the environment and of the finite planet imposing limits on economic growth led me to the conviction that my profession would never be the same again.

Surely there was no sense in perfecting the very methods and approaches that had brought us to this point. Intellectually, I became lost, having no idea how to take the first step in a new direction, or any idea what that direction ought to be. Even the most elementary question stumped me. How was I to know, for example, when specifying that a particular part or component be made out of aluminum or plastic, whether I was contributing to or subtracting from the environmental crisis? My professors had no

answers to such questions either, and so I came to the conclusion that, for my postdoctoral work, I had better find some answers.

This work led to my encounter with professional fundamentalism. The methods and approaches that were good enough for our professional predecessors were apparently still good enough for us. We engineers were busy with minute portions of the advancing technical frontier—so busy, in fact, that we did not appear to notice that the world around us was changing substantially.

Despite ongoing rhetoric about professional ethics, paramount responsibility to the public interest, and a need for sustainable development, the fundamentals of engineering methods and approaches were not being questioned. The content of engineering textbooks remained qualitatively unchanged. Of course, they were full of newer and better methods and approaches, and computers created new possibilities too. However, even the most casual inspection of the indexes in these books for references to human life, society, ecosystems, or the biosphere revealed that, if they were mentioned at all, they had no significant influence on the subject matter at hand. I could only conclude that we, as engineers, were in the grip of a professional fundamentalism—a metaphor I chose deliberately.

Another surprise awaited me when, at the postdoctoral level, I continued my studies of technology via the social sciences and humanities. When I entered the intellectual worlds of disciplines such as economics, sociology, political science, psychology, and ethics, I discovered them to be comparatively empty of technology. I appeared to be caught between incommensurate intellectual worlds. My own professional world was full of technology, leaving little room for human beings, societies, and the biosphere. The intellectual worlds of the social sciences and humanities, on the other hand, rarely included technology. This begged a question. How clearly can we hope to understand our world if we don't consider it in the context of science and technology? Many disciplines reflect this apparent intellectual imbalance. It's as if a group of artists sat side by side to paint a particular landscape, but each artist rearranged the scene by putting different things in the fore-

Consider pollutants. We first create them, then we install control devices to remove the most dangerous ones from waste streams. Then we landfill them. This merely transfers the pollutants from one medium to another without solving the initial problem.

ground and the background. Anyone unfamiliar with the original scene would have no idea what the landscape really looked like, or what aspects originally predominated. Are most disciplines "painting" technology too small, and out of proportion with everything else? Is our civilization in this predicament with respect to our understanding of reality?

This intellectual and professional soul-searching gradually led to a coherent research agenda. At its core lies the concept of preventive approaches for the engineering, management, and regulation of modern technology. It may be regarded as the prescription that stems from the following diagnosis.

The engineering, management, and regulation of technology, which powers much economic growth, have many practitioners making decisions. These decisions have consequences that usually fall outside the practitioners' domains of expertise. The undesired consequences of these decisions can be addressed only by other specialists who are competent to deal with them. This produces a system that first creates problems and then solves them, making it next to impossible to find the root of any problem in order to prevent or minimize it. As a consequence, the system displaces problems rather than solving them. It feeds on its own mistakes, and it traps us in what I have called the "labyrinth of technology."

To illustrate this, consider pollutants. We first create them, then we install control devices to remove the most dangerous ones from waste streams. Then we landfill them. This merely transfers the pollutants from one medium to another without solving the initial problem. We are feverishly restructuring corporations to improve workplace productivity, which, as shown by socio-epidemiology, has made work one of the primary sources of physical and mental illness. This, in turn, is necessitating additional and expensive health and social services. But these remedies do not go to the root of the problem either, and can only grow to the detriment of corporations, workers, and society. We keep adding devices and services to the system, but they do not fix it. It is even more distressing that if we subtract the costs incurred in the creation of wealth from the gross wealth produced, economists are beginning to recognize that the results are radically different from what the gross domestic product shows. Since the late 1970s, there has actually been a reduction in the net wealth produced. As the GDP kept growing, most of the population became poorer, and the quality of life was steadily undermined by a variety of social and environmental problems. The "signal" of the desired consequences of technological and economic growth is increasingly challenged and undermined by the "noise" of the undesired consequences.

In my search for answers, I needed to delve more deeply into the inner workings of technology and the economy. First, I carried out a comprehensive study of undergraduate engineering education to obtain answers to the following questions: How much do future engineers learn about how their design and decision-making affect human life, society, and the biosphere? Do they learn to make use of this understanding to adjust their design and decision-making to prevent or greatly reduce undesired and harmful effects?

These questions were converted into research instruments that were extensively tested. They were then used to quantitatively score each component of the curriculum, including textbooks, course notes, student lecture notes, laboratory manuals, assignments, tests, and examinations. These instruments also included supporting materials, such as videos, or first-hand experiences, such as field trips. It took my research team two years to complete the job and find an answer. The discovery? Future engineers learn almost nothing about these issues.

We then investigated the possibility that engineers learn this type of design and decision-making on the job when they become exposed to specific technologies in specific areas of application. We investigated six of these areas: materials and production, energy, work, the built habitat, computer-based systems, and technology transfer for development. We used the research instruments to

score the methods and approaches in these six areas. We also used them to determine to what extent engineers anticipated consequences in order to use this understanding to adjust design and decision-making to prevent or minimize undesired and harmful effects.

Once again, the research team found that the practice of addressing such potential problems at their root was virtually absent. The success of the methods and approaches used in the above-listed areas of application was assessed in terms including efficiency, productivity, profitability, and cost–benefit ratios. This measured the ratio of desired outputs to requisite inputs of materials, energy, labour, capital, and specialized knowledge. However, these ratios provided no information about whether any gains were genuine, or if they had been achieved by degrading human life, society, and the biosphere. The exception was a very small cluster of methods and approaches pioneered in industry. Apparently, the way engineers learn to steer technology into the future is like learning to drive a car only by watching the dashboard instruments, and never looking out the windows. It is little wonder that gross wealth production is being undermined by the costs incurred in producing that wealth. Put another way, as the GDP rises, people and ecosystems decline.

Preventive approaches for the engineering, management, and regulation of modern technology require an understanding of what I call the "ecology of technology." This shows how modern technology is embedded in, dependent on, and interacting with human life, society, and the biosphere.

Preventive approaches for the engineering, management, and regulation of modern technology require an understanding of what I call the "ecology of technology." This shows how modern technology is embedded in, dependent on, and interacting with human life, society, and the biosphere. I have constructed such an ecology based on two principles. The first principle governs the flow of materials and energy, and is derived from the first and second laws of thermodynamics. The second principle arises from the observation that the socio-cultural world derives its integrality from the way we live both individual and collective lives.

With the aid of this "ecology of technology," specialists participating in the engineering, management, and regulation of modern technology can trace the consequences of decisions that fall beyond their domain of competence. Next, design and decision-making can be adjusted by using this understanding in a negative feedback mode to prevent, or greatly reduce, harmful and undesired effects. It turns out that the most innovative and economical practices in industry essentially conform to this pattern. We have documented this in six areas of application: materials and production, energy, work, cities, computer-based systems, and technology transfer for development. By fitting the developments in industry into the conceptual framework of preventive approaches, it has been possible to develop an overall preventive strategy that would allow the Canadian economy to produce goods and services more competitively, while substantially lowering social and environmental burdens. This confirms what was already common sense: It is much cheaper to prevent the problem in the first place than to produce it and then have to deal with it.

Perhaps the primary barrier toward implementing a preventive strategy is neither technical nor economic. It is a scepticism among professionals that the solution looks like a free lunch and is, therefore, too good to be true. This simply reflects their difficulty in thinking beyond end-of-pipe approaches.

And where are we now? The Centre for Technology and Social Development is encouraging other engineering schools to become partners in opening up this new frontier in engineering theory and practice. It is also developing a program to transfer this knowledge to Canadian industry. The centre's research findings are being documented and will comprise eight volumes of material. Finally, the centre offers courses at the graduate and undergraduate levels that deal with preventive engineering and social development, with twelve hundred students taking at least one course per academic year.

Our evidence suggests that a nation that succeeds in systematically exploiting the potential of preventive approaches will become a leader in economic competitiveness, in reducing the harmful

effects that modern ways of life impose on society, and in making its way of life much more sustainable. I hope Canada will be the nation to benefit from this important innovation.

Further Reading

The Labyrinth of Technology, Willem H. Vanderburg (Toronto: University of Toronto Press, 2000).

Material Concerns: Pollution, Profit and Quality of Life, T. Jackson (New York: Routledge, 1996).

Factor Four: Doubling Wealth—Halving Resource Use, E. Von Weiszäcker, A.B. Lovins, and L.H. Lovins (London: Earthscan Publications, 1997).

Healthy Work: Stress, Productivity and the Reconstruction of Working Life, R. Karasek and T. Theorell (New York: Basic Books, 1990).

The Death and Life of Great American Cities, J. Jacobs (New York: Modern Library, 1993).

With an Honours B.Sc. in physics and an M.Sc. in electrical engineering, both from Université Laval, *Dr. Sylvain Houle* is well suited for the multidisciplinary aspect of positron emission tomography (PET) imaging. Also, as the director of the Vivian M. Rakoff PET Imaging Centre at the Centre for Addiction and Mental Health and the University of Toronto, he finds himself perfectly positioned to pursue his broad research interests, which range from PET imaging technology, to the development of PET radioligands and their application in brain disorders.

While at Université Laval, Dr. Houle first became interested in applying methods from physics and engineering to medical research. He pursued this interest further at the University of Toronto, where he enrolled in electrical engineering as a Ph.D student. His research at the Biomedical Research Institute involved a new tomographic imaging technique for nuclear medicine. However, still wanting to pursue a career in medical research, he felt he should learn more about medicine itself and enrolled in the medical school at the University of Toronto.

Having retained his interest in medical imaging, Dr. Houle specialized in nuclear medicine after completing medical school. Shortly after graduation, he became head of the division of nuclear medicine at the Toronto General Hospital. In the early 1990s, he was selected as director for the PET facility planned for the Clarke Institute of Psychiatry (now part of the Centre for Addiction and Mental Health).

Imaging the Troubled Mind

Sylvain Houle

For a group of classmates at Université Laval in 1969, it was an evening like many others. Sitting in the student lounge discussing the day's lectures. Facts. Figures. Equations. And an abundance of inspiring, yet mind-bending, ideas.

We had learned how an electron and its antimatter companion, the positron, can annihilate each other and transform their mass into pure energy. The result of this is two high-energy photons, and a perfect example of Einstein's famous $E = mc^2$, we were told. Very nice, but nothing to do with us. It was an abstract physics concept that we would remember for a future exam. At least that's what we *thought*.

Much higher on our list of concerns that evening was the occasionally bizarre behaviour of one of our classmates. At times, he seemed to lose touch with reality, and he was becoming paranoid about professors "wanting to fail him." A bit overboard, we thought. In our ignorance, we blamed his increasingly unusual behaviour on the stress of our upcoming final exams. At least that's what we *thought*.

At the time, I had no way of knowing how wrong our *thoughts* had been on both counts. I couldn't have known that, many years later, these two separate events would converge and take on such significance in my professional life.

Looking back with the experience and knowledge I have gained since those university years, I now realize that the brain of our highly intelligent classmate was becoming overwhelmed by the disorderly thinking of schizophrenia. I also realize that, even though we couldn't find much use for those theories about positrons and their inevitable annihilation, there were others in the research world who were paying close attention. Fortunately, they were doing something about it.

In fact, physicists and engineers were already hard at work building an imaging device that would use the properties of positron annihilation to explore the human brain, and to provide us with the tools we needed to explore the terrible disorders that can affect the mind. The device would become known as the positron emission tomography (PET) scanner, the latest of a series of instruments that rely on radiation to probe the chemistry of the human body. Many years later, when I assumed responsibility at the PET centre at the Clarke Institute, a facility devoted to imaging the troubled mind, it was hard not to see the irony. Things had come full circle.

It was, perhaps, fitting that some of the early PET experiments at the Clarke Institute—now part of the Centre for Addiction and Mental Health—involved probing the brain's mechanism of memory. It had been known for many years that brain activity induces changes in local blood flow and metabolism in those areas of the brain participating in a particular mental task. Some of the pioneering work in this area had been done in Scandinavia, using radioactive gases and a helmet of radiation detectors placed around the subject's head. The advent of PET allowed researchers in Canada to expand and refine those early experiments.

As radioactive nuclei decay, they emit high-energy photons, known as gamma rays. These rays can be sensed by radiation detec-

Neurotransmission can be altered by drugs—leading both to possible treatments and to addiction. While there are many different types of neurotransmitters, two of them have been the main targets of PET research: dopamine and, more recently, serotonin. These two brain chemicals play a central role in Parkinson's disease, schizophrenia, depression, and addiction.

tors placed outside the body. When they were first developed, the detectors were unable to localize the distribution of those radioactive substances within the body. Over several decades, however, advances in electronic, computer, and radiation detector technologies have allowed scientists to build instruments that are now able to image the distribution of that radioactivity in the living human body. As a result, a new field of medical imaging, called nuclear medicine, was born. It relied on minute quantities of radioactive substances, injected into a human subject. These substances would then participate in the body's physiological processes (without disrupting them) and provide an accurate measure of the body's chemistry.

Another major breakthrough occurred when the mathematical techniques used in astronomy were applied to medical imaging. At the core of these techniques was the back-projection algorithm, which led to the development of the computerized axial tomography (CAT) scanner. This was the first medical imaging device to use the new algorithm for producing three-dimensional X-ray images of the body.

Nuclear imaging and tomographic reconstruction, combined with the special properties of the positron, led to the development of positron emission tomography imaging. PET works because certain radioactive atoms decay by emitting a positron that rapidly annihilates an electron—resulting in two high-energy photons that leave the site of the decay in opposite directions. These basic facts of physics—conservation of momentum and Einstein's famous $E = mc^2$—were no longer abstract concepts learned in the classroom. They had become the keys to investigating the inner workings of the human brain and mind.

The focus of brain PET research is the chemical transmission occurring between brain cells. Brain cells communicate with each other using electrical signals carried along their axons. However, at the junction of two brain cells, the synapse (the electrical signal) is carried across the synaptic gap by chemical messengers. This process, known as "chemical neurotransmission," is often dysfunctional in people with a brain disease.

Neurotransmission can be altered by drugs—leading both to possible treatments and to addiction. While there are many different types of neurotransmitters, two of them have been the main targets of PET research: dopamine and, more recently, serotonin. These two brain chemicals play a central role in Parkinson's disease, schizophrenia, depression, and addiction.

Despite the major importance of brain chemistry in mental disorders, the application of PET to psychiatric diseases has lagged significantly behind that of its application to Parkinson's disease. By the time PET had become a practical reality, the role of the dopamine system in schizophrenia and its treatment had been well established. This was due, in part, to the contributions of Toronto scientists Oleh Hornykiewicz and Philip Seeman. In addition, PET studies at the Johns Hopkins Medical Institution in Baltimore, and at the Karolinska Institutet in Stockholm, had demonstrated not only that the dopamine system could be imaged with PET, but also that the effect of antipsychotic therapy could be measured quantitatively. Despite these dramatic demonstrations, the potential of PET in psychiatry remained largely ignored. Fortunately, Dr. Vivian Rakoff, then chair of psychiatry at the University of Toronto and president of the Clarke Institute of Psychiatry, realized the immense potential of PET in psychiatry. He spearheaded an effort to create the first PET facility in the world to be located in a psychiatric institute, and primarily dedicated to psychiatric and other mind disorders. I was honoured to be appointed founding director of the new facility. My task of actually building the facility and assembling its multidisciplinary team was made much easier thanks to the exceptional talents of my chief radiochemist, Dr. Alan Wilson.

Our initial work built on the long tradition of dopamine research in Toronto, and was centred on the role of dopamine in schizophrenia. Dr. Shitij Kapur was the first of a new breed of young psychiatrists, trained at the PET centre, to fulfill Dr. Rakoff's vision of PET as a bridge between basic and clinical research. Over the past several years, Dr. Kapur has worked with PET scientists and clinicians to understand and optimize the treatment of schizophrenia patients with antipsychotic drugs. As had been the case with the research on memory, the results of those PET studies shattered some long-established beliefs.

It had been customary to start treating patients who were newly diagnosed with schizophrenia with relatively high doses of antipsychotic medication, and to keep increasing the dose if they did not respond to treatment. Unfortunately, those antipsychotic drugs have unpleasant and serious side effects. Patients often discontinue their medication because of the side effects, leaving themselves vulnerable to the ravages of the disease. By using PET to measure the number of dopamine receptors occupied by the antipsychotic medication, it was shown that a good response to treatment could be obtained with much lower initial medication doses. This dose reduction had the added benefit of markedly reducing the occurrence of side effects. What is more, those same PET studies demonstrated that it was futile to increase the dose beyond the level where most of the dopamine receptors were already occupied by the medication. More recent work has revealed many new facts about the neurochemical effects of those medications. Before PET, it had been impossible to study those effects of drug treatment in the living brain of patients with a mental illness.

PET also had another impact on patients and their families. Even in our supposedly enlightened society, there is still a strong stigma attached to mental disorders. By revealing abnormalities in images of the brain, PET has demonstrated that mental illnesses and addictions are, in fact, true disorders of the human brain and mind.

Depression has not been as well studied with PET as schizophrenia because of the lack of suitable PET imaging agents. However, our group recently achieved a major breakthrough in that area when Dr. Wilson was able to develop an effective radiotracer to study the serotonin reuptake system with PET. After being released in the synaptic gap, serotonin is taken back into the neuron by specialized reuptake sites on the cell surface. These reuptake sites are the targets for medications that are widely used to treat depression, and are known as serotonin selective reuptake inhibitors (SSRIs). A young psychiatrist in our group, Dr. Jeffrey Meyer, is now using this new PET tool to study patients treated with SSRIs. Already, new discoveries about these medications are being made, discoveries that would not have been possible without PET.

PET also had another impact on patients and their families. Even in our supposedly enlightened society, there is still a strong stigma attached to mental disorders. By revealing abnormalities in images of the brain, PET has demonstrated that mental illnesses and addictions are, in fact, true disorders of the human brain and mind. If a picture is worth a thousand words, then, by using powerful images, PET has provided powerful new evidence. It is very gratifying to those of us involved in this research to see PET combating ancient misconceptions about psychiatric illnesses.

As the new millennium unfolds and my memory recalls those events of years past, I realize that I could never have foreseen how the disconnected events of that particular evening—sitting around the student lounge with classmates—would one day be meshed together. I was not able to predict the future then, and have learned not to try to do it now.

There are some things that are certain, though. In the coming years, PET will play an increasing role in both the understanding of existing treatments and the development of better drugs. Together, they will help to ease the suffering of those with mental illnesses and of their families.

As professionals, our task will be made easier by the funding received from the Canada Foundation for Innovation, and by the matching funding from the Ontario Innovation Trust. That support will allow us to acquire the most advanced PET scanner yet built to probe the neurochemistry of the human brain.

Further Reading

Mind and Brain: Readings from Scientific American Magazine (New York: W.H. Freeman and Co., 1993).

Molecules of the Mind: The Brave New Science of Molecular Psychology, Joe Franklin (New York: Dell Publishing, 1987).

Images of the Mind, Michael I. Posner and Marcus E. Raichle (New York: Scientific American Library, 1994).

Searching of Memory: The Brain, the Mind, and the Past, Daniel L. Schacter (New York: Basic Books, 1996).

Fifteen years ago, *Dr. Robert Wolkow* was another example of the brain drain, but today he is back home in Canada and flourishing.

In 1987, Dr. Wolkow was lured away to the IBM research lab in New York to work on a brand-new scanning tunneling microscopy technique. The job required a scientific jack-of-all-trades—someone with skills in electronics, a little software development experience, and an understanding of surface physics and chemistry. At IBM, Dr. Wolkow found himself fascinated by the microscopy technique, and soon made enhancements to it that allowed the first-ever atom-by-atom view of a surface chemical reaction. The work led to a classic research paper and an award for the most outstanding work of the year at IBM.

Dr. Wolkow later took up a staff position at the legendary Bell laboratories (birthplace of the transistor), where he made the first tunable, low-temperature scanning tunnelling microscope (STM). His machine, operational after three tough years of development, allowed him to solve the structure of the silicon surface, a problem of more than thirty years' standing. As Bell labs turned away from basic research, Wolkow and his family came home to Canada and a position at the National Research Council—where he would start a scanning tunneling microscopy group.

Dr. Wolkow was born and raised in Hamilton, Ontario. He received a B.Sc. from the University of Waterloo, and a Ph.D. from the University of Toronto. He has twice received the National Research Council's highest achievement award and holds the rank of principal research officer. Recently, he received the Noranda Award and the Rutherford Medal for physics from the Institute of Physics, and has been elected a Fellow of the Royal Society of Canada. He is considered a leading figure in the new field of nano-scale science.

Molecular Devices: The Next Technological Revolution

Robert A. Wolkow

Imagine a grain of rice sitting on a kilometre-long stretch of highway. For all intents and purposes, that's the size of a molecule compared to a centimetre. For many years, scientists have imagined schemes for creating devices out of these tiny molecules. The smallest imaginable structures, they would be built of only a few molecules but would have the power and potential to support powerful computers, new communications tools, and fantastic medical diagnostic devices. Those who thought it was all just science fiction are now realizing that the ideas were based on sound scientific principles, and much of the thinking stands up today.

Unfortunately for those early molecular dreamers, one seemingly insurmountable problem stood in their path: No one could think of a way to actually *make* the proposed molecular units, let alone apply them to the latest technology. To appreciate how big the problem was, we have to stop for a moment to think small—and contemplate the scale of the elements involved. A molecule is about one billionth of a metre in length, which is just about the relative size of that grain of rice on the highway. Even compared to the microscopically small transistors that make up today's electronic devices, molecules are very tiny things.

In recent years, two issues have refocused attention on molecule-based technologies. One is imminent need, the other is a new capability. As the brightest researchers chart the journey of semiconductor device technology on their "technology roadmap" to the future, they have come to realize there is a formidable roadblock up ahead. Experts predict that in ten to fifteen years there won't be any more room to grow—or, in this case, to *shrink*. At that point, they fear that most semiconductor devices will be as small as physically possible and will have lost their cutting-edge patina. They will have become mere commodities, in turn causing a weakening of the world economy. As a result, scientists and industry are looking for the next technology, and molecules are high on the list of potential candidates. Aside from the fact that they are very, very small, molecules are attractive because they have richly varied properties, which give them the potential for a wide range of uses. The idea is to use molecules to make new devices that will break through the roadblocks on the trip to future technology, and allow for the production of better, cheaper, and faster devices.

The force that propels us toward new technology is clear, but what about the technological challenges? How easy is it to fabricate molecular devices? In the past several years, some extraordinary developments and key advancements have made it possible to start realizing the dream of manipulating matter on the molecular scale. The 1981 invention of the scanning tunneling microscope (STM) by Gerhard Binnig and Heinrich Rohrer at the IBM labs in Switzerland was precisely that kind of development.

The STM is a deceptively simple tool and not at all like a conventional microscope. It doesn't use lenses and there is no eyepiece to look through. In fact, the STM is much closer to a record player with the sharpest needle imaginable—at its tip, only as wide as a single atom. Under computer control, the tip gently

scans over a surface, detecting every tiny hollow or minuscule bump. As it moves across the surface, a topographical map appears on the computer screen—point by point, line by line. Within minutes, a detailed map of a surface is complete. The importance of the STM is widely recognized, and in 1986, Binnig and Rohrer were awarded a Nobel Prize for their work.

In the mid-1980s, the basic STM technology was transferred from Zurich to the IBM Research Center in Yorktown Heights, New York. Some very smart scientists in the New York lab seized on the STM technique and quickly made adaptations and improvements. Soon after that, I was hired to work at the IBM lab. I rushed there early in 1987, even before defending my thesis or graduating.

Scientists and industry are looking for the next technology, and molecules are high on the list of potential candidates.

I was instantly fascinated with the STM technique. The excitement I had over its potential was coupled with long hours, intense focus, frequent setbacks, and occasional bursts of discovery. After having added a few new features to the STM, I had my own success: the recording of the first-ever atom-by-atom view of a surface chemical reaction.

Today, the STM is causing a revolution in science. It has become a powerful, enabling, and widespread tool. Scientists around the world have contributed to the steady advance of its technique and to its increasingly diverse applications. As a result, problems that were utterly impossible to solve when I was a student are now routinely handled. Mature areas of study have been rejuvenated, and new fields have emerged.

The excitement has spilled over into the popular press in the form of fantastic claims about the coming age of molecular machines or nanotechnology. Unfortunately, most of the reports have been misleading. Indeed, the combination of intense interest in molecule-based technology and the improbability of realizing any actual dividends in the very near term poses a real threat. There is a danger that enthusiasm, including financial backing, will dry up before we realize the true potential of new possibilities. To avoid building unrealistic expectations, we must encourage more discourse between scientists and the public.

Which raises a related issue. Does a revolution in science automatically translate into a revolution in technology? The short-term view suggests no, but the practical impact of scientific upheavals is more or less assured. It's just that the connection between a scientific breakthrough and a tangible product might be convoluted and take a long time to make.

Quantum mechanics is a good example of this long road from theory to practice. Nearly a hundred years old, the theory has often been portrayed as an abstract and esoteric construct. Yet it has actually led to practical, world-changing applications. The same can be said for spectroscopy, a field pioneered by Canadian Nobel laureate Gerhard Herzberg. One of my favourite examples is the nuclear magnetic resonance (NMR) machine. When it was first developed in a physics lab fifty years ago, no one imagined that it would be the foundation for the lifesaving magnetic resonance imaging (MRI) machines commonly used in hospitals today.

Are we on the verge of yet another technological revolution? We have a long way to go, but it seems clear that the emergence of powerful molecular devices is inevitable. The first applications may be only five or ten years away, but it will likely take decades before revolutionary new molecule-based devices are commonplace. It seems we are at a similar point to that just before the transistor emerged from Bell labs in the 1940s. At the time, the underlying physics for transistor technology was on fairly firm ground, and many scientists believed the development of a solid-state electron valve was inevitable. Indeed, the transistor was a project goal, not an accident. Just as transistors struggled for a decade or so before integrated circuits emerged, so it may be that molecular devices will have to fight to be competitive before advanced fabrication techniques allow them to find their stride.

But the transistor analogy only goes so far. Transistors were up against vacuum tubes, which were a functionally excellent, but bulky, expensive, and energy-gobbling competitor. In contrast,

It doesn't make sense for molecules to compete with silicon as a computing engine. For now, they can't. Just as we can easily predict the outcome if a novice fighter steps into the ring with a champion, we can do the same in the show-down between silicon and molecules. To gain a foothold, molecules need a niche. So why not aim at silicon's weaknesses?

replacing silicon with molecular technology will be a formidable task, largely because silicon has many near-ideal properties. For example, silicon's conductivity can be tuned over a very wide range—essentially from insulating to metal-like—simply by adjusting the concentration of dopant atoms in crystal. It also has the advantage of a naturally forming oxide with excellent insulating and passivating properties. Combined with sophisticated lithographic techniques, these properties make it possible to sculpt very fine, and numerous, electrical paths and switches into a silicon crystal, thereby creating the powerful devices that surround us today. That's a hard act for any molecule to follow.

Since great technical challenges remain, and silicon enjoys towering strength, we are faced with one burning question: How do we move forward from here? It doesn't make sense for molecules to compete with silicon as a computing engine. For now, they can't. Just as we can easily predict the outcome if a novice fighter steps into the ring with a champion, we can do the same in the show-down between silicon and molecules. To gain a foothold, molecules need a niche. So why not aim at silicon's weaknesses?

Silicon-based technology is not suitable for light emission or light detection. This makes it impotent in modern optical communications applications. Moreover, silicon cannot begin to interact with molecules in the infinitely varied and sensitive ways that they interact with each other.

I suggest that we build molecular devices on a silicon platform. This type of silicon-molecule hybrid strategy helps get around the extraordinarily challenging problems related to "wiring up" a molecular device. Just as importantly, a hybrid strategy can be aimed at *enhancing* the capabilities of silicon rather than competing with it head-on. This approach aims to marry silicon with the natural capacity of molecules to absorb and emit light, and to share in the subtle and discriminating interactions that life processes depend on. As well, the hybrid approach opens the door to unlimited new applications in molecular sensing and medical diagnostics.

Year by year, progress continues. Scientists have recently discovered ways to controllably move individual atoms, creating fantastic new structures and demonstrating a key capability for making functional molecular entities. At the National Research Council of Canada, we have developed techniques that allow us to determine exactly how organic molecules can attach to silicon surfaces. This is a key step toward our goal of creating hybrid silicon-organic devices. Much work remains to be done before practical hybrid silicon-molecule devices can be built, but, more than ever, it seems likely that the necessary fabrication procedures will be developed and that applications will be realized. Once we get over the hurdles, hybrid devices could be used in a wide range of areas, such as telecommunications, medical diagnostics, prosthetic devices, and probably others we cannot yet imagine.

Can Canada benefit from this opportunity? Yes, if we invest now. With a renewed scientific infrastructure and an injection of several hundred million dollars over five years, we can establish a leading presence in this field. We can also ensure substantial dividends, capture key intellectual property, and grow new industries in the years to come.

Further Reading

Crystal Fire: The Invention of the Transistor and the Birth of the Information Age, Michael Riordan and Lillian Hoddeson, Sloan Technology Series (New York: W.W. Norton, 1998).

Designing the Molecular World, Philip Ball (Princeton, NJ: Princeton University Press, 1994). Winner of the 1994 Association of American Publishers Award for Best Professional/Scholarly Book in Chemistry.

"Molecular Level Fabrication Techniques and Molecular Electronic Devices," F.L. Carter, *Journal of Vacuum Science and Technology*, Appendix B 1 (1982), pp. 959-68.

"Controlled Molecular Adsorption on Si: Laying a Foundation for Molecular Devices," R.A. Wolkow, *Annual Review of Physical Chemistry*, Vol. 50 (1999), pp. 413–41.

Dr. David Schindler is a Killam Memorial Professor of Ecology at the University of Alberta. From 1968 to 1989, he founded and directed the Experimental Lakes Project of the Canadian Department of Fisheries and Oceans near Kenora, Ontario. In his work with the project, he conducted interdisciplinary research on the effects of eutrophication, acid rain, radioactive elements, and climate change on boreal ecosystems. His work has been widely used in formulating ecological management policy in Canada, the U.S., and Europe.

Dr. Schindler received his Ph.D. from Oxford University, where he studied as a Rhodes Scholar. He has headed the International Joint Commission's Expert Committee on Ecology and Geochemistry, and the U.S. Academy of Sciences' Committee on the Atmosphere and the Biosphere. He has served as the president of the American Society of Limnology and Oceanography, and as a Canadian National Representative to the International Limnological Society.

An author of more than two hundred scientific publications, Dr. Schindler has received numerous research awards, honours, and distinctions. He is a Fellow of the Royal Society of Canada, has served as a member of the Royal Society's Global Change Committee, and has received four honorary degrees. In August 1991, he was presented with the Stockholm Water Prize by the Queen of Sweden for his research on the acidification and eutrophication of lakes.

His more recent awards include the Manning Award of Distinction for innovation in science in Canada (1993), the first Romanowski Medal of the Royal Society of Canada (1994), and the Walter Bean–Canada Trust Award for Environmental Science (1996). In October 1998, he was a co-recipient of the Volvo Environment Prize for his significant contribution to insights on how chemical pollutants and microbiological factors affect the quality of fresh water. In 1999, he received the J. Gordan Kaplan Award for Excellence in Research from the University of Alberta, and the ASTech (Alberta Science and Technology) Award for Outstanding Leadership in Alberta Science. In December 2000, Dr. Schindler received an Award of Excellence from the Natural Sciences and Engineering Research Council (NSERC). Most recently, Dr. Schindler received NSERC's 2001 Gerhard Herzberg Canada Gold Medal for Science and Engineering, in recognition of his role as a world-leading environmentalist, and for highlighting the importance of science in guiding decisions about the natural world.

The Combined Effects of Climate Warming and Other Human Activities on Canadian Fresh Waters

David Schindler

As Canadians, we are used to having abundant supplies of fresh water in our lakes and rivers. Despite its importance to life on earth, however, fresh water has been our most mistreated and neglected natural resource. In fact, Canadians have already abused their fresh waters in many ways.

In the coming century, climate warming will exacerbate the effects of many human activities and degrade Canadian fresh water on a scale that is incomprehensible to the average Canadian. Quantity and quality are just two of the many problems that will make freshwater protection one of the most important economic and environmental issues of the twenty-first century.

On average, Canadians consume about 326 litres of water per day at home—about twice the per capita water use of Europeans, and many times that of people in Middle Eastern countries. This figure does not include the water used for industrial, agricultural, or hydroelectric power purposes, which would greatly increase the estimated water use figure. Population growth, industrialization, expansion of agriculture, increased demand for hydroelectric power, and other activities will greatly increase the demand for fresh water in the years ahead.

At first glance, it might seem that Canada's freshwater sources are capable of meeting its demand. Lakes cover 7.6 percent of the country's surface—over 755,000 square kilometres. Canadian rivers discharge nine percent of the global freshwater flow. Fourteen percent of our land surface is covered by wetlands, while a further two percent lies beneath snow and ice. There are also extensive groundwater reserves in most areas.

However, we often forget how much of our water is unavailable for consumption. Most Canadian rivers flow northward, away from the three-hundred-kilometre-wide band along the United States border where almost all of the country's thirty million people reside. Much of the fresh water in western Canada originates in the snow and ice fields of the Rocky Mountains, which are both receding and thinning as a result of global warming. Though most global climate models do not predict large changes in precipitation for Canada, evaporation increases rapidly as climates warm—drying small streams and lowering water levels in lakes. This would bear serious consequences for shipping and navigation. Climate models predict that by the end of this century, average temperatures in Canada will be several degrees warmer than they were in the middle of the twentieth century. During the warmest period of the past ten thousand years, known as the mid-Holocene period, average annual temperatures were only one to two degrees Celsius warmer than they have been in the mid-twentieth century. Scientific studies show that the mid-Holocene period was disastrous for fresh waters.

The growth of human population, industry, and climate warming will increase the demand for Canadian water by water-poor regions. The burgeoning populations and water-usage habits of the American Southwest have already caused many of their rivers and aquifers to be oversubscribed. By the time it reaches the Gulf of California, the mighty Colorado River has been reduced to a trickle. The Oglala

aquifer, which serves much of the west-central regions of the United States, is being exploited eight times faster than its waters are being renewed by natural sources. These conditions will become worse as climates warm, freshwater sources dry out, and the demand for Canadian water increases. The export of our water would generate enormous profits, which would become irresistible to politicians and industrialists. Unless we act now to prevent it, the situation will be the same as the one we now face with natural gas. In short, the fate of our waters will be determined by international demands.

Increased evaporation resulting from climate warming will exacerbate many pollution problems, because as water flows decline, the concentrations of pollutants increase proportionately.

Until recently, most Canadians assumed that our fresh waters were of high quality. The Walkerton tragedy, and the subsequent revelations about deficiencies in water treatment, have provided something of a wake-up call. Bad management of the watersheds of lakes and rivers has exacerbated water-quality problems. Erosion allows silt, fertilizers, pesticides, and pathogens to be swept directly into watercourses. The draining of wetlands, and the deforestation of the riparian zones that line the banks of streams, further diminish natural protective mechanisms for aquatic ecosystems. Overuse and bad application procedures for manure and other fertilizers are also common.

Increased evaporation resulting from climate warming will exacerbate many pollution problems, because as water flows decline, the concentrations of pollutants increase proportionately. Only vigilant protection of our watercourses will ensure that we have high-quality water that's good enough to drink and to support aquatic life in the next century.

My passion for ecosystems first began with an interest in biology. It dates back to my boyhood in northern Minnesota, where I spent all my spare time exploring lakes and forests. After I had made my interest in biology known to all around me, I was advised that a biologist could be either a high school teacher, which was abhorrent to a sixteen-year-old, or a medical doctor, which was a full-time indoor profession. So I decided to put my math and science skills to good use by enrolling in engineering physics at the University of Minnesota. At the end of my freshman year, I found a summer job in the biology department at North Dakota State University. My boss, Gabe Comita, whose enthusiasm for science and biology was contagious, was a limnologist. It was a profession I'd never heard of. After I found out that limnology involved the study of the physical and chemical properties of bodies of fresh water, and the conditions of their plant and animal life, I found myself just as enthusiastic. I switched to biology at North Dakota State that fall. My plan was to become a limnologist.

After I completed my doctorate at Oxford University, I accepted a position in Canada as an assistant professor at Trent University, just north of Peterborough, Ontario, where a wide variety of lakes were situated in different geological settings. I was then offered and accepted a job at the new Freshwater Institute in Winnipeg, administered by the Fisheries Research Board of Canada. In my new position, I would be the director of a field station where whole-lake experiments could be done to investigate the scientific underpinnings of policies for sound management of fisheries and fresh waters. The project would be located in a remote part of northwestern Ontario. My duties that first summer were to locate a site for the field station, get a road built, select fifty lakes to preserve for our experiments, and design the first experiments. I had found the job of my dreams after all. I formed the Experimental Lakes Area (ELA) to do science as I believed it should be done.

A year later, we began our first experiments to test the roles of phosphorus, nitrogen, and carbon in the eutrophication (overfertilization) of lakes. At the time, a fierce debate was raging between academic scientists and the detergent industry's scientists and officials. The academics believed that phosphorus was the element that had to be controlled. Those in the detergent industry believed that nitrogen or carbon management would also be needed. Since detergents contained high concentrations of phosphorus, industry

In 1973, I tried to convince senior bureaucrats that acid rain was a major threat to Canadian lakes. The acidity of rain and the geological setting in Eastern Canada were similar to those in Scandinavia, where damage caused by acid rain had already been documented. The bureaucrats, however, regarded acid rain as a small problem and were not willing to provide further funding for the research.

officials were reluctant to change their formulations. Our lake experiments quickly showed that controlling the carbon or nitrogen that resulted from human activities was futile since the atmosphere naturally resupplied lakes with these elements. We were also able to clearly show that phosphorus control would reduce the eutrophication problem in most lakes. New legislation followed that limited the phosphate content of detergents, and required removal of phosphorus from sewage effluents in the Great Lakes basin. The policies were successful and were soon copied in other countries.

While all of this was happening, the Fisheries Research Board was disbanded, and we were placed within the civil service; first in Environment Canada, then in Fisheries and Oceans. Since we had solved the eutrophication problem so quickly, government bureaucrats reasoned that the ELA should be closed and staff assigned to other problems. I, however, believed that experiments in whole ecosystems were a vital part of environmental science.

In 1973, I tried to convince senior bureaucrats that acid rain was a major threat to Canadian lakes. The acidity of rain and the geological setting in Eastern Canada were similar to those in Scandinavia, where damage caused by acid rain had already been documented. The bureaucrats, however, regarded acid rain as a small problem and were not willing to provide further funding for the research. At just about this time, we had a stroke of luck that would revitalize the ELA and the government's commitment to it. The Alberta tar sands were opening, and there was great interest in how vulnerable the soft-water lakes might be to acid rain. Along with the renewed interest came the funding for new experiments, and three years later, there was new evidence that acid rain was a problem in eastern Canada. The government released money for our research.

Our studies of the acid rain present in lakes also produced several previously unknown discoveries. The potential damage of acid rain had been determined largely by exposing fish for ninety-six hours in tanks of water at various levels of acidity. The results indicated that damage to fisheries was not likely to occur at pH values above 5 (pH is a chemical symbol used to express the acid or alkaline content in water and soils. A value of pH 7 is considered neutral; pH 0 denotes high acid content; pH 14 denotes high alkaline content). In fact, few lakes in North America were acidic enough to have pH values below 5.

There may have been plenty of fish in our "sea," but our research showed that there were also many other organisms lower in the food chain. These organisms were more sensitive to acidification than fish were. Some species that supported lake trout were eliminated at pH 6, ten times less acidic than pH 5. With the organisms eliminated, and nothing to eat, lake trout began to starve. Eventually they ceased to reproduce at pH values of around 5.5. By the time pH values were decreased to 5, where damage was thought to begin, our lakes had lost one third of their biodiversity. No species of fish were reproducing, and several key ecosystem functions had been impaired. We were also able to show that microbial processes in lakes would assist in their recovery if acid deposition were decreased. It had earlier been believed that once acidified, lakes would never recover. The results and benefits of our research reached far beyond our lakes. They helped to persuade policy-makers in several countries that it was necessary to curb the emissions of sulfur oxides that caused acid precipitation.

Looking back on our work at ELA, there are many people who contributed to our success who deserve recognition. Wally Broecker, of Columbia University's Lamont-Doherty Earth Observatory in Palisades, New York, was an important partner in many of the experiments at ELA. His insights into geochemistry were an excellent complement to my own in biology, and he was able to obtain money from the United States National Science Foundation to help fund whole-lake experiments. Several students from the observatory did their Ph.D. work at ELA, and then went on to become eminent limnologists, oceanographers, and geochemists. By discovering and quantifying some of the key geochemical processes that were changed as a result of our experiments, Wally and his students were an important part of what made ELA successful.

Today, as a professor at the University of Alberta, my enthusiasm is renewed, and I intend to spend the remaining years of my career teaching new scientists to perform research and to fight fiercely for the ecosystems that support us.

During the entire period of study, we also monitored the physics, chemistry, and biology of a number of reference lakes and streams in the ELA. This enabled us to better interpret the results of our experimental perturbations. In addition to serving as a reference, these lakes allowed us to measure the effect on fresh waters of a relentless twenty-year period of climate warming—when average air temperatures increased by almost two degrees Celsius between 1970 and 1990. The data provided us with a preview of how lakes would be affected by climate warming. Increasingly, forest fires are denuding watersheds, exposing lakes to more wind. Harmful UV radiation has penetrated deeper into lakes as inflow streams have dried up. This is cutting off the stained organic matter that has provided a natural sunscreen for aquatic organisms. The oxidation of sulfur compounds in shallow lake sediments and peatlands has accelerated acidification. The habitat for cold-loving fish and the general productivity of the lakes has declined.

In 1989, I left the ELA, disheartened by these changes to the ecosystems that I loved. In addition to climate warming, clear-cut logging was beginning to occur throughout the area, and the ELA eventually became a small green oasis in a sea of destruction. Reduced funding also made it almost impossible to do large-scale experiments. Today, as a professor at the University of Alberta, my enthusiasm is renewed, and I intend to spend the remaining years of my career teaching new scientists to perform research and to fight fiercely for the ecosystems that support us.

Now, perhaps more than ever, we must fight. If we are to maintain the integrity of freshwater ecosystems and prevent the collapse of freshwater fisheries, then individuals and communities must practise water conservation. They must also prevent livestock and the runoff of fertilizers and pesticides from fouling our waters. To do this, they must change their usage and preserve or restore the streams and lakeshore areas that have been modified by human activities. Also, in areas where there is a variety of human activities, basin-wide management plans are needed to protect the integrity of our waters.

We should also restore our commitment to the type of research and monitoring programs that allowed us to avert or reduce many freshwater problems in the 1960s and 1970s, which included overfertilization of lakes with nutrients, and acid rain. As well, we need to make faster progress in international negotiations to reduce emissions of greenhouse gases and toxic contaminants. If we are to protect the fresh waters that are so important to Canadians, we must mount a comprehensive program of public education, research, monitoring, and enforcement. The current approach, which relies on recovery after problems have been caused, and which treats problems in isolation, is far too simplistic to compete with the multifaceted attack on our freshwater systems by industrialized human society.

Further Reading

"Eutrophication and Recovery in Experimental Lakes: Implications for Lake Management," D.W. Schindler, *Science*, Vol. 184 (1974), pp. 897–99.

"Long-Term Ecosystem Stress: The Effects of Years of Experimental Acidification on a Small Lake," D.W. Schindler et al., *Science*, Vol. 228 (1985), pp. 1395–1401.

"A Dim Future for Boreal Waters and Landscapes: Cumulative Effects of Climate Warming, Stratospheric Ozone Depletion, Acid Precipitation and Other Human Activities," D.W. Schindler, *Bioscience*, Vol. 48 (1998), pp. 157–64.

"The Cumulative Effects of Climate Warming and Other Human Stresses on Canadian Freshwaters in the New Millennium," D.W. Schindler, *Canadian Journal of Fisheries and Aquatic Sciences*, Vol. 58 (2001), pp. 18–29.

Recognized internationally for his work in genomics, *Dr. Thomas J. Hudson* is director of the Montreal Genome Centre, and past assistant director of the Whitehead Institute/MIT Center for Genome Research.

At the Massachusetts Institute of Technology (MIT) in the early 1990s, Dr. Hudson led the effort to generate dense physical and gene maps of the human and mouse genomes. With his MIT experience and accomplishments in hand, he came back to Canada and, in June 1996, founded the Montreal Genome Centre, based at the McGill University Health Centre Research Institute. He has also played a pivotal role in launching Genome Canada and Genome Quebec.

Widely regarded as a leader in the development and application of robotic systems and DNA chip-based methodologies for genome research, Dr. Hudson's interests in human genetics focus on the dissection of complex genetic diseases. Ongoing disease projects in his laboratory include the search for genes predisposing people to lupus, inflammatory bowel disease, coronary artery disease, asthma, and diabetes. His laboratory is also using DNA chip technology to characterize breast and ovarian cancer.

Dr. Hudson teaches in the departments of human genetics and medicine at McGill University, and practises medicine at the McGill University Health Centre of the Montreal General Hospital (Division of Immunology and Allergy). He has received numerous awards during his career, including the Clinician Scientist Award from the Canadian Institutes of Health Research. His work also includes various private-sector consulting activities.

Genomics: A Question of Scale

Thomas J. Hudson

At first, not many scientists shared our excitement about the Human Genome Project. Considering where we stand today, it's hard to believe, but true.

The project was deemed expensive. Wouldn't the money be better spent on other projects? Would we be spending scarce funds to study what was mostly junk DNA? Perhaps the most negative comment made to a young scientist interested in finding his way through the human genome was that there was "no science" involved with the project. Thankfully, this pessimism did not prevail at the Massachusetts Institute of Technology (MIT), where my career took shape. Instead, the early attitude was a mixture of confidence, a touch of arrogance, and determination.

I joined the Human Genome Project in 1991, more by a series of chance events than by design. An early event occurred when Montreal clinician-scientists Emil Skamene, Joe Shuster, and Phil Gold each brought me to their offices to recommend that I jump-start my research career by interrupting my clinical work in Montreal and plunging into molecular biology. With their support, I landed at MIT. It was an opportunity that I was eager to tell my dad about, since he too had done his post-graduate training at MIT. The MIT experience was both daunting and fascinating. The talent that surrounded me was impressive. David Housman, my first director, was a pioneer in human gene discovery, and his laboratory had almost two dozen projects cooking, each led by talented Ph.D. scientists.

Another event that led me to genomics seemed more like bad luck at first. My arrival in Cambridge had been happily delayed by the birth of my second child, Tommy. By the time I made it to MIT, the project that had been planned for me had been started by another lab member. I spent the next few weeks restlessly roaming about a busy laboratory, waiting for a project. In time, David brought me together with Nic Dracopoli, who was setting up the second genome laboratory at MIT. We called it "Genome South" to distinguish it from Eric Lander's laboratory, "Genome North."

My work with Nic was supposed to be temporary. I was expected to generate some human DNA markers for "chromosome 1," using the resources of the genome labs, which were dedicated to generating maps of mouse chromosomes. There was no glamour in the project. The work was extremely repetitive. I was either hand-pipetting small solutions from one tube to another, or streaking thousands of bacteria with toothpicks. Most of my work was done after four o'clock, since the lab was too crowded during the day to accommodate me and my tasks.

A restless mind wanders easily during long hours spent picking bacteria from one agar plate and streaking them onto another plate, or pipetting solutions from one tube to another. I wondered, couldn't the process be automated? Would it not eliminate mistakes and be faster? Perhaps robots could work through the night, and then the poor postdoctoral scientists could go home early.

I assume it was my determination to overcome such limitations that led Eric Lander to invite me to become group leader of the human genome mapping group at the Whitehead Institute, at MIT's Center for Genome Research. In the following decade, this centre

A restless mind wanders easily during long hours spent picking bacteria from one agar plate and streaking them onto another plate, or pipetting solutions from one tube to another. I wondered, couldn't the process be automated? Would it not eliminate mistakes and be faster? Perhaps robots could work through the night, and then the poor postdoctoral scientists could go home early.

would establish its leadership in genomics through its development of high-throughput DNA technologies. This would be followed by early delivery of its promised products: maps and sequences of human, mouse, and other chromosomes. I was one of the project participants, having led the group that developed a robot called the Genomatron, a monster instrument that generated three hundred thousand DNA tests a day. The genome team used it to generate high-resolution maps of all human and mouse chromosomes—early and critical milestones on the path to sequencing these genomes.

The Genome Project started as an ambitious attempt to identify and classify all the genes of an organism. It made a lot of sense to start there, since genes are the basic building blocks of life. Although daunting by their size, genomes are relatively simple problems (since they exist as linear molecules) compared to the subsequent processes to be tackled.

Before we continue along the genome map, though, let's take a moment to consider some important elements of genomics. Supporting and sustaining the genome map is some very fundamental molecular biology.

DNA → RNA → Protein

In order to function, genes must first be transcribed to ribonucleic acid (RNA), which is then translated into protein. Since this dogma was first postulated—soon after the discovery of the structure of deoxyribonucleic acid (DNA) in 1953 by Watson and Crick—

biologists have been able to characterize many basic biological processes on a molecular scale by carefully analysing the individual components of these pathways. The expanded dogma is also a simplification of biology:

Genome → Genes → RNA → Proteins → Cell → Tissue → Organ → Organism → Population → Ecosystem

In reality, biology studies processes of life, which are complex for many reasons, including size (large genomes contain over thirty thousand genes), interactions (biological processes occur in cells expressing thousands of proteins interacting in multiple ways), and diversity (individuals within the same species usually differ because of hundreds of thousands of discrete changes in their genomes, in conjunction with diverse environmental conditions).

As a result, life is possible because of an enormous number of complex processes, all happening simultaneously. Genomics (and proteomics—the global study of proteins in a cell) does not propose changes to the central dogma, or to the ever-present knowledge that life is complex. Genomics is really a question of scale. It uses large-scale approaches to understand life.

It was this understanding of life that all of us at MIT were pursuing, but it would take more than confidence, obstinacy, ideas, luck, and a little arrogance to accomplish our job. It would take essential financial resources. Funding for this work at MIT, which came mostly from the United States National Institutes of Health, provided for equipment, scientific personnel, administrative personnel, reagents, and space. Thankfully, some healthy risk-taking was thrown in—such as the $1.2-million financing of the Genomatron. Attention to support personnel was also important in order to protect the "science" time of lead scientists. Space crises were rapidly resolved, with four major expansions in the decade following the beginnings of Genome North and Genome South. Flexibility was given to the lead scientists to modify the research program (and equipment lists) to be as competitive as possible.

In 1996, I began weekly travel between Boston and Montreal with the flicker of an idea: The Montreal Genome Centre. For the

time being, this was an idea without resources. Not surprisingly, it was Skamene, Shuster, and Gold who convinced the Montreal General Hospital Foundation to donate funds for two automated sequencers and a pipetting robot. This donation came at a time when support for genome research in Canada was at its lowest. "Too expensive," was what we heard. "Scarce resources," "The Americans can do it all," and so on. Nevertheless, confidence did exist among a number of key people in Montreal, such as Kenneth Morgan, a colleague and co-founder of the genome centre; Daniel Gaudet, a key collaborator on several genetics projects; Abe Fuks, dean of medicine at McGill; and Gervais Dionne and his colleagues at BioChem Pharma. The support of the latter group was unexpected at the time, since I didn't know them. I remember them, though, as the first of many groups in Montreal that expressed the strong sentiment that we can do things here just as well as anywhere else.

As the human genome sequencing project is being completed researchers are now dealing with the next big challenge, that is, how to study global processes, from single cells to entire populations.

The Montreal Genome Centre became a reality in 1999, when we received major infrastructure funds from the Canada Foundation for Innovation, the Quebec government, BioChem Pharma, Applied Biosystems, and others. The genome group expanded its technology base from high-throughput genotyping (a process used in localizing disease genes), to sequencing and DNA chips. However, it is of great importance to me that we have gone beyond the implementation of existing technologies, and started developing new technologies, particularly in the area of DNA chips. The group is expanding rapidly. Applicants for postdoctoral positions abound, and come from all over the world.

The Montreal Genome Centre can rightfully claim to be one of the first genome centres to specialize in medical genomics, a term used to describe the applications of genome information, resources, and technologies that are used to understand, diagnose, prevent, and treat human diseases. Clinically relevant information will come, slowly but surely.

The scope of disease processes studied at the Montreal centre spans a spectrum of common diseases including asthma, multiple sclerosis, heart disease, and ovarian cancer. More than one hundred Canadian investigators, as well as groups from the U.S., Israel, and New Zealand, have requested the centre's expertise in the past two years. In addition to working together with Hôpital Ste-Justine to co-identify a gene causing a childhood neurological disease frequently found in Eastern Quebec, the centre has localized new genes that are implicated in adult diabetes, cataract, cirrhosis, and epilepsy. Some of the lead projects study drugs with multiple-disease relevance. This includes corticosteroids, which are useful in the treatment of inflammatory diseases such as lupus and asthma, but which have such harmful side effects as osteoporosis, diabetes, and high blood pressure.

In general, though, what a genomic approach offers to medical research is a macroscopic view. Whereas molecular biology has developed great tools to study genes one at a time (a so-called reductionist approach), new genomic technologies, such as DNA chips, allow all the genes in a cell to be studied simultaneously. That is the difference. This global approach allows scientists to see how a disease process unfolds, and then to witness new gene pathways. More and more, we identify similar patterns among subsets of disease samples, leading to new classes of disease (for example, a particular tumour is categorized into two classes, each with different responses to medication).

As common diseases such as hypertension, cancer, and asthma are understood to represent mosaics of different disease processes, we can expect to see the emergence of clinical tools that will help us to classify the diseases that affect our patients more specifically. This will be followed by therapeutic approaches that are tailored to the patient. This transition from bench to bedside will be complex for health care providers, patients, and society, but armed with careful clinical studies and responsible legislation, society can fully profit from the genomics revolution.

As the human genome sequencing project is being completed researchers are now dealing with the next big challenge, that is,

how to study global processes, from single cells to entire populations. Biologists are joined in this endeavour by engineers, mathematicians, chemists, computer scientists, physicians, ethicists, demographers, and sociologists. The impact of genomics can be seen in a number of areas, including the high-throughput DNA, RNA, and protein technologies currently in development; the advent of bioinformatics and voluminous biological databases; the tremendous growth of the biotechnology industry; and the weekly media stories of disease-gene discoveries that the public greets with great interest.

And they thought we were studying junk DNA.

Further Reading

"The Business of the Human Genome," Carol Ezzell, *Scientific American* (July 2000), p. 40.

"The Human Genome Business Today," Kathryn Brown, *Scientific American* (July 2000), pp. 50–63.

"Beyond the Human Genome," Carol Ezzell, *Scientific American* (July 2000), pp. 64–69.

"The Race Is Over," F. Golden and M.D. Lemonick, *Time* (July 3, 2000), pp. 13–24.

Human Genome Project information Web site:
http://www.ornl.gov/TechResources/Human_Genome/home.html

Index